Global Ecology

Global Ecology

Colin Tudge

Oxford University Press
New York 1991

Designed by Janet McCallum

Planned and produced by
The Natural History Museum
Cromwell Road, London SW7 5BD

Published in the United States of America by
Oxford University Press, Inc.,
200 Madison Avenue,
New York, N.Y.   10016

Oxford is a registered trademark of
Oxford University Press

Library of Congress
Cataloging-in-Publication Data

   Tudge, Colin.
     Global Ecology/Colin Tudge.
        p.   cm.
     Includes index.
     ISBN 0-19-520904-4
     1. Ecology.   I. Title.
     QH541.T819 1991   91-2090
     304.2--dc 20   CIP

ISBN 0-19-520904-4

Printing (last digit) 9 8 7 6 5 4 3 2 1

Typeset by Tradespools Ltd, Frome, Somerset
Printed in Italy by Arnoldo Mondadori, Verona

Contents

Foreword

The word 'ecology' seems to have been in the language forever. In fact it was coined just over a hundred years ago, in 1873, and is derived from the same Greek word – *oikos*, meaning house or dwelling – as the word 'economy'. It is the study of how living things – plants, animals, fungi, bacteria, viruses – interact with their physical environment and with each other. As biologists now calculate that there may be anything from 10 to 50 or more *million* different species of living creatures on Earth, and as the physical environment is more intricate than we can ever measure, ecology has emerged as an infinitely complex pursuit.

For decades – through most of the first half of this century – the word 'ecology' seemed merely to be a piece of academic stuffiness: a smart name for natural history. Now the word 'ecology' is on everyone's lips. It features in political manifestos; no political speech is complete without it. It's what children want to learn about at school, and everyone wants to study at university. It is the fit and obvious subject for The Natural History Museum to present, with its long tradition of bringing the best and most relevant science out of the laboratory, and making it accessible. So what has changed? Why has this word, 'ecology', an obscure piece of scientific Greek, burst so dramatically onto the public stage?

In the past few decades the subject has matured: become broader, deeper, and more robust. Without losing sight of the beauty of nature, and the sense of wonder that nature should engender, the science has become more rounded and satisfying. It grades into the study of animal behaviour (ethology) and of evolution; and embraces the modern notion of 'survival strategy'. It of course involves the study of living mechanisms (physiology). The physical properties of the environment that must be taken into account now include almost everything that a physicist and chemist might contemplate: the radiations from the Sun and stars, the intricate surface properties of clays, the interactions of stratospheric gases, as well as the more earthbound measurements of temperature, moisture, pressure, and the rest. Underpinning modern ecology too – as is the case with every respectable science – is mathematics; indeed, some of the most striking advances in ecology in the past few years have been made by mathematicians. We will not dwell on maths in this book; but it's nice to know it's there, as a constant supporter and tester of ideas. To be sure, ecology can never be as precise as physics: living nature is too complex, and too subject to the vagaries of climate, ever to be exactly predictable. But ecology is a proper science, now, and repays proper involvement.

Also, everyone now can see that the study of ecology is *relevant*. To be sure, as long ago as the Middle Ages, people complained about the spoiling of the environment. But the damage they perceived in those days was local: too much smoke in the cities, for example. Outside, they felt, nature was vast, 'untamed'. It could take any amount of punishment; whatever it had to offer was there for the taking. Many an essayist and poet railed against the despoliation of the planet as the industrial revolution got underway, at the end of the eighteenth century and the beginning of the nineteenth: Rousseau, Blake, Byron. But still the feeling persisted, well into this century, that there was always plenty of wilderness left, and that we could take whatever was there with impunity.

Now it is clear to everyone with eyes to see and with senses to feel that this just is not so. Everyone sees their own immediate environment degraded, it seems, by the day; and with it, the decline of the creatures that once lived there. More broadly, everyone is becoming aware of less visible but more far reaching forces: the warming of the planet through the greenhouse effect; the death of European trees and the sterilization of northern lakes by acid rain; the general feeling – which has long been true, but is only now being widely appreciated – that there is no pristine wilderness left, that everywhere has felt the impact of humanity, that there is nowhere left to retreat to.

The tragedy is that this realization is not yet translated into policy. Economic policies are still based on the ancient notion that it is physically possible for all human beings, or at least for vast numbers, to be materially rich; even though it should be obvious by now that the trappings of wealth, the metals and timber and fuels, are extracted at too high a price. Papua New Guinea is still regarded in the most influential circles not as one of the world's last great forests, but as a potential source of gold; even though gold is less beautiful than trees, and its extraction is horrendously polluting. It is clear, too, that almost all environmental problems can be made worse by yet further increases in human population. Yet our population seems bound almost to double again before it stabilizes, as very few countries have policies for birth control, and some regard such control as blasphemy.

The hope, though, comes from the growing awareness, among people at large, that the planet is in trouble, and that a change of ways is necessary; and from a growing appreciation of the beauty and value of species – an appreciation that seems to grow as animals and plants become more threatened.

And with that awareness, and appreciation, has come the realization that it isn't good enough simply to regret the despoliation of the Earth, and the passing of our fellow creatures. We have to act; and if we are to act sensibly, and do good, then we have to understand as best we can how the planet works and how its creatures survive; and, as physicians say, what ails the patient.

Appreciation of the planet comes from within us. Policies that may enable us to care for it must be made by governments. But the understanding that will enable us to do the right things, if we have the will to do them – that depends on the science of ecology. That is why it is important. That is why The Natural History Museum has mounted its major *Ecology* exhibition.

This book is an informal introduction to the principal ideas.

Colin Tudge

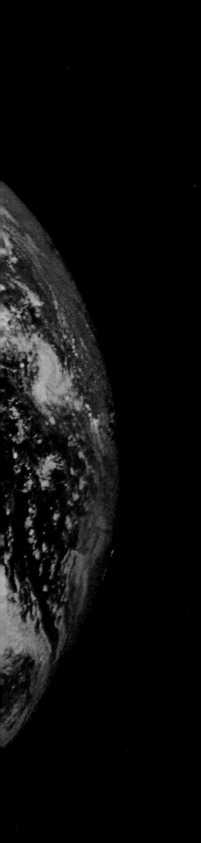

1 *Conditions for life*

The Universe is a hostile place. Most of it is far too cold and empty to sustain the intricate and fragile processes of life. The stars that punctuate the void contain far too much energy. Some places are so hot, indeed, that the atoms of which matter is made are torn apart; others so dense, that atoms collapse.

Only by the merest chance would a traveller through space happen upon the planet Earth; a tiny jewel of a place, orbiting an unexceptional star in an ordinary galaxy, but – miracle of miracles – providing precisely those conditions, neither too hot nor too cold, nor too dense nor too sparse, where the astonishingly delicate processes of life can unfold.

All this begs many questions. What is this elusive quality called 'life'? What forms does it take? Why is it delicate? Why does it require such peculiar conditions? Why are those conditions so rarely achieved? What is so very special about the planet Earth?

What is life?

The question seems easy enough, and so do the answers. Living things grow, move, reproduce, and respond to their surroundings, and non-living things do not. No-one could confuse a dog with a stone.

All these observations are true, and crucial, but they do not quite seem to get to the root of things. Crystals grow, but we do not feel that salt is alive. Air moves. Stones fragment. Metals and rocks expand and contract with heat and cold, and that is a response of a kind; certainly, the Universe is a turbulent place even in the absence of living things. Somehow to grasp the essence of *life* we must probe more deeply.

Three other qualities, less obvious, emerge when we look deeper; all of them relevant to the principal theme of this book.

The first is the idea of the *organism*. A crystal grows, to be sure, but each part of a crystal is simply a repetition of every other part; the same monotonous pattern of atoms, repeated over and over. A dog, though, is not a repeated pattern of parts of a dog; it is not a collection of ears or an assemblage of tails. Each part of the dog (broadly speaking) performs a different function; each of the many millions of different kinds of molecule in its billions of cells, contributes to the overall entity. A puppy is not a fragment of a dog; it is a near facsimile

1

▲ *Crystals have one of the characteristics of living things; they grow. But they lack the three essentials: they do not evolve, they do not process energy, and they cannot be called organisms.*

of the original, as complete, and with as many interdependent parts, as its parents. Intricate though it is, the organism is the unit, with all its many parts contributing to the whole. There is nothing like an organism in the non-living world.

The second is the notion of *energy*. Inanimate objects and systems slavishly obey Newton's law of thermodynamics that states that the energy in any system tends simply to dissipate; or, to put it simply, that 'things cool down'. Living things do not disobey these physical laws – but they *look* as if they do. They scavenge energy from all around them – from the rays of the Sun, or by consuming energy-rich molecules created by other living things; and they use that energy, constantly, to maintain their own extraordinary complexity. If ever they lose the ability to scavenge energy, and put it to use in repairing their own fabric, then they are said to be dead; and they begin immediately to fall apart. Much of ecology is about this process: finding energy; putting it to use.

The third notion is that of *natural selection* leading to *adaptation*. This is the key principle of modern biology, as described by Charles Darwin in the mid-nineteenth century. He observed that when living things reproduce, the sibling offspring vary. He pointed out, too, that in a world in which many organisms are reproducing, there is bound to be competition for resources. Because the individual organisms vary, some are bound to be better able to survive in particular circumstances than others. In general, those that survive are the *fittest* – in the sense of most apt, not in the modern sense of healthiest; in

▶ *Chance plays a large part in shaping evolution. In Madagascar, lemurs evolved to fill all the niches that are occupied elsewhere by monkeys and apes. But they were able to do so only because – by chance – monkeys and apes failed to reach this island.*

other words, those that are best *adapted*. The process by which the fittest are singled out for survival is natural selection.

It seems to follow from Darwin's ideas that as the generations pass, so the organisms in a particular line of descent (lineage) must become better and better adapted to the prevailing conditions. And indeed, there are many lineages in nature that seem to show this steady improvement in adaptation; modern horses are more beautifully fitted to running on the open plain than ancient ones; modern monkeys more agile than their predecessors.

If natural selection were given its head, and an infinite amount of time to work in, then it seems that all lineages would tend inexorably towards 'perfection'; where perfection was defined as perfect adaptation. In practice, life does not work out quite like that. For one thing, each lineage is limited in the things it can do. As the great British biologist/mathematician J.B.S. Haldane commented, human beings would be hard pressed, however long natural selection acted upon them, to sprout the wings of an angel; we just do not have the genes that would even begin to provide the appropriate structures, and past selection has never acted to provide us with these structures.

Of equal significance is that the rules of the game have a tendency to change. That is, physical conditions on Earth are prone to alter. Thus it was that the dinosaurs were astonishingly successful for 140 million years – a period of dominance in which the mammals of the day (and there were plenty of mammals about) were pushed into the background, or at least into holes in the ground and in trees. But the climate changed about 65 million years ago (perhaps exacerbated by a bombardment of meteors) and the dinosaurs, for reasons that are still unexplained, were unable to cope with the shift. Thus their various and superb lineages came to a halt.

It is also now very clear that *chance* plays an enormous part in determining which creatures survive, and where. It was chance, for example, that determined that the lemurs found their way into Madagascar before that island had drifted too far away from Africa, while the monkeys and apes did not; and so the lemurs of Madagascar had *carte blanche* to fill all the ecological niches that monkeys and apes came to occupy elsewhere. For these reasons – internal constraint, changing circumstance, and chance – natural selection does not lead inexorably towards 'perfection', or (apparently) to any predestined goal. Darwin knew these things perfectly well, and drew attention to them. Evolutionary theory may seem to belong more properly in a different book; but no subject in modern biology, and certainly not ecology, can be considered properly without keeping evolutionary principles in mind.

Professor Graham Cairns-Smith of the University of Glasgow has suggested that the propensity of lineages to adapt through natural selection is the most fundamental of all the qualities of living things. He has pointed out, too, that if life is considered in this light, then some clays could be thought of as alive. The particular clays he has in mind do exist in many different crystalline forms. In different circumstances, some forms are more likely to persist than others; in a stream, for example, some shapes are more likely than others to escape being washed away. Hence in any one set of circumstances some forms, rather than others, will be selected. This is an intriguing notion in many ways; and it may well be, indeed, that evolving clays played a part in the evolution of the creatures that we are more inclined to recognize as living. Clays do, however, lack the other qualities that we may consider fundamental. There are no clay organisms; and clays do not scavenge and process energy from their surroundings to keep themselves intact.

The idea of the organism; the notion that organisms need to process energy to stay intact; the appreciation that selection is constantly at work, favouring the best adapted – these are the fundamentals to keep in mind in all studies of ecology.

We may acknowledge, *pace* Professor Cairns-Smith, that some clays have at least one of the crucial qualities of living things. Nevertheless, the creatures we generally think are alive are not clays; they are dogs and humans and trees and mosses. What is so special about these, that they should have come to prevail?

The special properties of organic life

The organisms that prevail – those that are selected for survival – will be those that grow and reproduce most abundantly. These will be the ones that scavenge energy from their particular surroundings most efficiently, and use it to repair and augment their own bodies, and to produce facsimiles of themselves.

The scavenging of energy and the processing thereof is a complicated business. It requires complicated chemistry. To do it well, indeed, requires the property we have identified in those organisms we generally recognize as living: a whole variety of different and complex molecules, all working in concert.

Molecules of the necessary complexity are not easy to create. There are only about one hundred different kinds of atoms in the Universe, each kind known as an *element*; and we know that this is the case because we know the fundamental laws of chemistry that determine the structure of atoms, and can see that only a certain number of structures are theoretically possible. Of all that hundred, only two are theoretically capable of producing the extremely intricate molecular structures that are needed to process energy in the way that is required. These are silicon and carbon. The reason they have this versatility is that they are able to join up with no fewer than four other atoms at a time, to form chains or networks. Most other atoms can join up comfortably only with one, two, or three others at a time.

Of the two, carbon – which has the smaller atoms – is by far the more versatile: it lends itself much more easily than silicon does to the creation of intricate molecular shapes. Clays, including the putatively 'living' clays, are made primarily of silicon. The creatures we conventionally think of as living are all carbon-based.

Perhaps, elsewhere in the Universe, there are other living things.

The principal compounds of life

Living things consist and make use of many millions of different kinds of organic (carbon-based) molecules. In practice, however, most of these these myriad compounds belong to one or other of just four classes of compound: carbohydrate, fat, protein, and nucleic acid.

Carbohydrates are the sugars, and the compounds compounded from sugars, such as starch, glycogen, and cellulose. The first three of these serve as forms of energy (starch mainly in plants, glycogen mainly in animals) while cellulose is the tough material from which the cell walls of plants are constructed. Thus, cotton consists of cellulose; and so does wood, though wood is toughened with lignin. Sugars also attach to proteins to form 'glycoproteins', and it now seems that most proteins function *only* in the form of glycoprotein. Carbohydrates in pure form contain only carbon, hydrogen, and oxygen.

Fats are extremely varied. They are sources of energy – indeed they contain twice as much energy per unit weight as either carbohydrate or protein. They serve as essential components of cell membranes. And – in the form of cholesterol – they form the basis of many animal hormones, particularly the steroid hormones, which include the sex hormones. Fats also consist exclusively of carbon, hydrogen, and oxygen.

Proteins have four main uses. They form much of the substance of living things: they are essential components of cell membranes, and of cell contents in general; and the muscles of animals are made mostly of protein. Secondly, and vitally, proteins are enzymes: the catalysts of living things that control all the mechanisms of life. Thirdly, in animals, some proteins serve as hormones, such as growth hormone. Finally, proteins may also serve as energy sources. Some wild animals, such as deer, rely heavily on protein – muscle tissue – as a prime source of energy in winter; and seeds contain storage proteins, such as gluten in wheat, hordein in barley and zein in maize. Proteins consist of chains of sub-units known as amino acids, each of which contains carbon, hydrogen, oxygen, and nitrogen. Sulphur is important too, as some common amino acids also contain small amounts of this element, and virtually all natural proteins contain some of these sulphur-containing amino acids.

Nucleic acids are of two main kinds – DNA and RNA. DNA is the stuff of which genes are made: the basic units of heredity which determine what kind of creatures develop. In most organisms, RNA serves as an assistant to DNA, but some viruses contain RNA *instead* of DNA. Nucleic acids contain carbon, hydrogen, oxygen, nitrogen, and phosphorus.

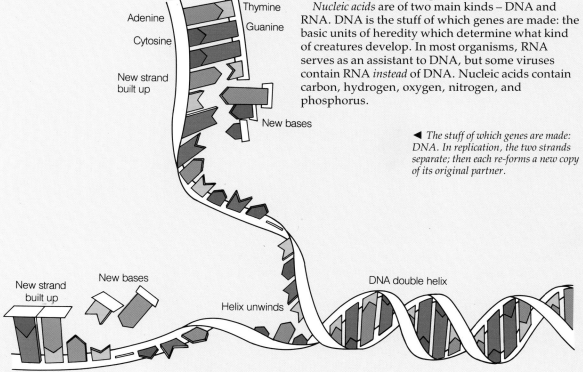

Adenine

Cytosine

Thymine

Guanine

New strand built up

New bases

◀ *The stuff of which genes are made: DNA. In replication, the two strands separate; then each re-forms a new copy of its original partner.*

New strand built up

New bases

Helix unwinds

DNA double helix

Perhaps there are quite different forms of life. Sir Fred Hoyle wrote a science fiction story in which a cloud of gas was intelligent, and anything that is intelligent has some claim at least to be considered 'alive'. So we should not be dogmatic, and assume that the kind of life we have on Earth is the only kind that could exist in the entire Universe; and 'living' clays may flourish elsewhere, perhaps even on Mars (as suggested by Professor Hyman Hartman of the Massachusetts Institute of Technology). On the whole, though, there are good chemical reasons for thinking that the most likely forms of life in other parts of the Universe would also be carbon-based, as on Earth.

In general, materials that are carbon-based are said to be *organic*. Organic chemistry is the chemistry of carbon. Biochemistry is the study of all molecules associated with life – most of which are carbon-based: so biochemistry is largely (though not exclusively) a refined branch of organic chemistry.

Carbon is versatile in chemistry in the way that bricks are versatile in architecture: it seems able to give rise to an infinity of different molecules, just as bricks can be employed to create a garden path or a palace. Indeed, the carbon-based molecules of which living things are constructed, and which they employ for life's purposes, are so various and can be so complex that the study of biochemistry might at first sight seem quite impossible. What makes it manageable is that the carbon-based compounds of which living things are made mainly belong to one of four classes: carbohydrates, fats, proteins, and nucleic acids. These are discussed on p.5. Living things also employ and contain plenty of other carbon-based materials that do not fit so easily into these neat compartments. These include lignin, which helps to give some of the strength to wood, and compounds such as alkaloids, which plants produce for purposes that are largely unknown, but which include the ability to repel insect pests.

Carbon, however, does not perform a solo turn in the creation of living things. Pure carbon, indeed, is merely graphite: or, in a somewhat fancier form, diamond. Many other elements play essential roles in the structure and maintenance of living things, so these other elements are also relevant to our theme.

The elements of life: supporting players

All the interesting organic molecules of which living things are compounded – including the carbohydrates, fats, proteins, and nucleic acids – are made not simply from carbon, but also from hydrogen and oxygen. Proteins and nucleic acids also contain nitrogen. Scores of other elements also play small but essential roles in keeping life's machinery running.

Besides carbon, hydrogen, oxygen, and nitrogen, two other non-metals are of outstanding significance in living things: sulphur (which among other things is a common component of proteins); and phosphorus (which – among many other roles – is an essential component of nucleic acids). The principal metals employed in living processes are calcium, iron, magnesium, sodium and potassium.

We are dwelling on this basic chemistry because it is highly relevant to the study of ecology. The places where living things live, the ways that they live and the relationships between them, crucially depend upon their ability to acquire the materials they need to keep themselves intact.

Sometimes organisms acquire the particular materials they need in the form of the element; oxygen, for example, is taken in largely in the form of oxygen gas. In most cases they do not; they take in the element they require already compounded into some other form. Thus plants take in nitrogen mainly in the form either of ammonia (NH_3) or nitrate (NO_3), while animals obtain their nitrogen mostly in the form of protein. But both can be said to have a basic requirement for nitrogen, which they incorporate into their own proteins and nucleic acids. When ecologists talk of plants needing iron, or animals needing sodium, this is shorthand; it means that the particular creature must seek out (or be rooted close to) some compound that contains the element in question.

It is because carbon-based molecules are necessary for the kind of life that we find on Earth, and because those molecules are of necessity complex, that life itself is fragile. It is a miracle indeed that in a hostile Universe this planet should provide conditions where such fragility can persist. How come?

The special qualities of planet Earth

Organisms of the kind that now live on Earth would die virtually instantly if they were transported beyond its protective cocoon. Indeed, many of the complex molecules of which they are composed would be liable to fall apart.

Most proteins, on which all the functions of Earthly life depend, are *denatured* – which essentially means 'cooked' – if their temperature is raised above about 45°C; hardly hotter than a hot bath. Then again, the internal chemistry of living things is restless. It depends upon mobility; molecules moving from one part of the structure to another. This essential motion cannot take place except in a liquid medium. Water, in practice, provides the medium; and water is liquid (at the kind of pressures that prevail on Earth) only if the temperature is between 0 and 100°C.

Most of the Universe is much colder than this. The temperature of space is about −270°C. A few places are far, far hotter: stars may be millions of degrees C. There are not many places in the Universe, in short, where the temperature is consistently between 0 and 100°C – let alone, between 0 and 45°C, which Earthly life forms prefer. The planets and moons that do remain within this range have to be just the right distance from the right size of star, and should also be spinning, otherwise the side facing the star becomes extremely hot, and the other side extremely cold.

Pluto. Unsuitable for organic life. Too small to retain permanent atmosphere. Far too cold.

Uranus. Unsuitable for organic life. No solid surface. Too cold.

Neptune. Unsuitable for organic life. No solid surface. Too cold.

Saturn. Unsuitable f
life. No solid surface

Whether or not organic life exists elsewhere in the Universe, it is bound to be rare. Of the nine known planets in the solar system, only Earth is known to support life. Mars could do so in theory, however, and Venus may have done so once, but is now far too hot because its atmosphere, rich in carbon dioxide, has produced a runaway greenhouse effect. In addition, three of the planetary moons — Europa, around Jupiter; Titan, around Saturn; and Triton, next to Neptune — could, perhaps, sustain some organic material. Not to scale.

Jupiter. Unsuitable for organic life. Made of gas — no solid surface. Too cold.

Mars. Could support life, but has not been shown to do so. Lacks an ozone layer, so ultraviolet could be too high for comfort.

Earth

c.
d.

Mercury. Unsuitable for organic life. Too hot. No atmosphere.

Venus. May have had life once. Now, runaway greenhouse effect makes it too hot.

The Earth is just the right distance from a star, the Sun, which is of just the right kind of size. The Earth also spins, quickly enough to ensure that the nights are not too cold, nor the days too hot.

A place for life to exist must also have liquid water on its surface; and in this, temperature is only one consideration. For if a planet or moon is too small then it will not have enough gravity to retain an atmosphere unless it constantly replenishes it by evaporation from the surface; the gases simply float away. If it has no atmosphere, then it will be surrounded by space, which is a near vacuum. If liquid water appeared on the surface of a world that was in a vacuum, it would evaporate instantly. We must count ourselves lucky, then, that our own Earth is big enough to retain an atmosphere, and therefore to retain liquid surface water.

The atmosphere has one other essential quality. Big organic molecules, and especially the nucleic acids of which genes are made (see p 5), are liable to be broken down by high-energy radiation, and in particular by the ultraviolet (UV) light that is a component of sunlight. The atmosphere filters out most of this radiation. In particular, a variant form of oxygen known as ozone, which exists mainly in the upper atmosphere of Earth, filters out most of the potentially dangerous UV light (see p. 89).

The Earth, then, provides just the right conditions for life to persist. The Universe contains so many billions of stars besides our own Sun, many of which have planets, of which many in turn have moons, that some of them are *bound* to provide conditions that could support organic life. We can see, though, that appropriate conditions must be rare. In our own solar system, only two planets out of nine provide conditions that could sustain life, and they are Earth and Mars. So, too, in theory, could Titan, one of the moons of Saturn.

We have inherited the present Earth. By the time human beings first came into being, several million years ago, life had already been evolving for three and a half *billion* years. But how did life begin?

The origin of life on Earth

Our Earth is believed to be about 4.5 billion (4500 million) years old, and life is believed to have begun on it around 3.5 billion years ago. Conditions, then, were in some respects similar to those of today; in some respects different, but in ways that do not matter; and in other respects different in ways that are very significant indeed.

An important similarity between the world of 3.5 billion years ago, and the present, is temperature. At its very inception, the world was hot, but within a billion years of its origin, parts of it at least were at the kinds of temperatures in which complex organic chemistry could take place.

Intriguing but unimportant differences were that the Earth spun much faster in those early days – so the days and nights were much shorter; and the moon was much closer, so it would have appeared vast in the sky (if there had been any creature to observe it) and the tides would have been far more dramatic.

▶ *Since the Earth began, 4.5 billion years ago, its atmosphere has changed dramatically. At first it was filled with gases that modern organisms would find highly toxic, such as ammonia and possibly hydrogen cyanide – though these are the gases that probably gave rise to organic life. But as photosynthesis (see p.93) evolved, so the atmosphere began to accumulate oxygen, and take on its modern composition.*

The important differences were in the atmosphere. Four-fifths of the modern atmosphere consists of nitrogen gas. Roughly one-fifth is oxygen. Then there are small but crucial traces of carbon dioxide, small and unimportant traces of rare gases such as neon, and traces of a whole range of other gases, some of which are very important. Modern living things are for the most part adapted to the modern atmosphere. Carbon dioxide is the ultimate source of carbon for all modern creatures, and most (but by no means all) creatures need oxygen for respiration.

The atmosphere of 3.5 billion years ago was quite different. It contained a whole range of gases that most modern organisms would find noxious, not to say highly toxic, which would indeed kill them within seconds. These included methane, ammonia, probably hydrogen cyanide, hydrogen sulphide and carbon monoxide. But the atmosphere of that time contained virtually no free oxygen (that is, oxygen gas).

How come? How could life have originated in an atmosphere which – by today's standards – seems so inimical to life?

Well, note first of all that the various gases that are thought to have made up the early atmosphere contain most of the main elements that are known to be essential to life: carbon, nitrogen, sulphur, hydrogen, and oxygen – though the oxygen was not 'free', but was

combined for example with carbon in carbon monoxide. All these gases are known to be highly chemically reactive. It is easy to imagine them combining together to form complex organic materials, of the kind that make up modern living tissue. Note, too, that because there was no free oxygen gas, there would have been no ozone layer – because ozone is a form of oxygen. So ultraviolet radiation from the Sun would have been felt more powerfully then, than now. This would have encouraged chemical interaction between the gases of the Earth's early atmosphere to form organic molecules – although, to be sure, it would also have tended to break those molecules apart once they had formed!

How come there was no oxygen, though? Why is there now? How did living things evolve in its absence?

The answer to the first question is simple. Oxygen is extremely reactive, and any that appeared in an atmosphere containing ammonia, hydrogen cyanide, and the rest, would immediately have reacted with them. And – in answer to the second question – the only reason that oxygen gas exists in such large amounts in the atmosphere today is that plants and some bacteria produce it in vast quantities through *photosynthesis*. This is the process by which these organisms synthesize the compounds of life from water and carbon dioxide, using energy from the Sun, and releasing oxygen (see p.93). Photosynthesis probably began to evolve about half a billion years after the first living things appeared – about three billion years ago – and so the process of oxygenating the atmosphere began. The existence of oxygen in the atmosphere is the most striking example of the fact that living things profoundly affect the fabric of the planet itself – for the atmosphere should be considered part of the fabric.

The answer to the last of the three questions should perhaps be put the other way round. Because oxygen is so reactive, it is potentially extremely toxic. It is oxygen that causes fats to go rancid. The membranes that surround the cells of *all* modern organisms are compounded of protein and fat, and oxygen given a free run at those membranes would be extremely destructive. Indeed the only reason that modern living things are able to survive in the presence of oxygen, is that they contain a variety of compounds that prevent it from reacting with materials such as fats: compounds that include vitamins C and E, and uric acid.

The fact that so many modern creatures utilize oxygen in respiration, and indeed now *rely* upon it (so that animals quickly suffocate in its absence), is a fine example of evolution adapting creatively. Because oxygen is so reactive, it can 'burn' organic molecules efficiently – breaking them down thoroughly, ultimately to form carbon dioxide and water. Organic molecules broken down in this way release virtually as much energy as they are capable of releasing – so such 'oxidative' breakdown is extremely efficient. Natural selection favours efficiency. So organisms that developed the ability to use oxygen in this way – that is, to practise *aerobic respiration* – quickly became dominant. We, of course, together with virtually all other animals, have inherited this ability, and respire very efficiently.

► *Until about two billion years ago, the dominant forms of life on Earth were bacteria. These often formed huge structures with mineral skeletons – similar to these stromatolites that still can be found on the west coast of Australia.*

▼ *Oxygen gas is not essential for all life. The first living things evolved in its absence, and many organisms even today still respire without its aid. Yeast is one such. It respires anaerobically, by a process known as fermentation; breaking down sugars to produce carbon dioxide and alcohol.*

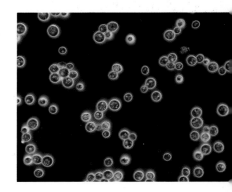

Three kinds of life: viruses, prokaryotes, and eukaryotes

Biologists divide living things into three broad categories: *viruses*, *prokaryotes*, and *eukaryotes*. To understand the basis of this division, we should look more closely at the construction of organisms.

All living things on Earth are composed of cells – sometimes just a single cell, but sometimes (as in dogs, humans, or oak trees) containing billions of cells. Each cell is a unit: it contains genes in the form of DNA, which determine what form the cell will take; and it contains machinery for protecting and for carrying out the 'instructions' of those genes. But the organization of the cell differs, fundamentally, between the three broad categories of organisms.

The cells of viruses are so simple, they hardly qualify as cells at all. They consist simply of a core of DNA, surrounded by a protective coat of protein. In some viruses, DNA is replaced by RNA. Indeed, virus 'cells' lack the essential machinery of life – such as enzymes for carrying out respiration and supplying energy; so viruses can 'live' and replicate only by invading the more 'complete' cells of bacteria, plants, or animals, and living as parasites within them. Because they live in this way, viruses are often *pathogenic* – they cause disease. As we will discuss later, pathogens in general – including viruses – are now known to be of enormous ecological importance (see p 132).

Although viruses have such a simple structure, biologists do not conclude that they are necessarily 'primitive'; that is, that they evolved a very long time ago as precursors of more complex organisms. It is just as likely that some viruses evolved from more complex organisms – that they are cell fragments which found they could earn a perfectly good living simply by sponging off others. Apparent simplicity in living things – as in man-made machinery – does not always spell primitiveness. It may imply refinement.

Prokaryotes are the bacteria, and related organisms. Their cells are far more complex than those of viruses, but they are still relatively simple. The term prokaryote is Greek for 'preliminary cell'. They contain sufficient biochemical machinery to sustain life, all surrounded by a 'proper' cell membrane made from protein and fat. But their genes – DNA – are simply distributed around the cytoplasm, the material that makes up the body of the cell. Some of the DNA is packaged into a giant chromosome, and some is contained within small packages known as plasmids. The genes of prokaryotes have a straightforward structure, consisting of uninterrupted strands of DNA,

whereas the genes of eukaryotes are interrupted by apparently nonsensical sequences of DNA known as introns. The reasons for this difference – and the reasons for the mysterious presence of the introns – is not understood. Whatever the cause, though, it seems to be of fundamental importance.

Bacteria are generally thought of as being single-celled. But many kinds of bacteria in nature form elaborate colonies, often quite visible to the naked eye, in which different individuals perform different functions, so that the whole colony functions as if it were a single organism.

Bacteria are extremely important ecologically in all kinds of ways. Some cause disease (that is, are pathogens), but most do not, and many are vital for the continued survival of all other living things. We will encounter them throughout this book.

Eukaryotes are the animals, plants, and fungi. Their cells have the most elaborate structure. The DNA is packaged into several chromosomes, and these in turn are contained within a nucleus – a distinct region with a membrane around it. In practice, the chromosomes are visible only during cell division, when the DNA of which they are mainly composed, contracts. But the nucleus itelf is clearly visible all the time, and the nucleus is indeed the hallmark of the eukaryotic cell. The term eukaryote is, in fact, Greek for proper cell.

Most of the organisms we see around us, and encounter directly, are eukaryotes. In the form of animals, trees, grasses and seaweeds, we could say that eukaryotes are the dominant organisms. But it is a mistake to think of ecology exclusively as an interplay of eukaryotes. Eukaryotes could not exist without bacteria, toiling away in the background; and the lives of all organisms, including bacteria, are profoundly affected by viruses. Ecology is the interplay of all three kinds of organism.

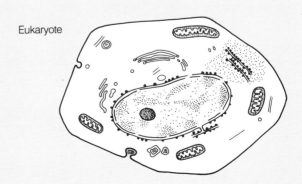

Eukaryote

Prokaryote

Three viruses

Herpes virus

Bacteriophage
(virus of bacteria)

Influenza virus

◀ ▲ *Three forms of life:
prokaryote, eukaryote and viruses.
Not to scale.*

However, many modern organisms respire without the aid of oxygen; that is, are *anaerobic*. Yeasts (which are fungi) are of this kind, and many bacteria. Indeed, some animal tissues still practise anaerobic respiration – including muscle fibres, for short periods. Most modern anaerobes (including most yeasts) are perfectly tolerant of oxygen, although they do not make use of it. But the world still contains ancient organisms that are not only anaerobic, but are actually killed by oxygen. These include the 'archaebacteria' that now live in airless marshes.

To return to our theme: how did the gases of the early atmosphere combine to form the organic compounds of which living things are made?

One leading idea is that gases such as hydrogen cyanide, methane, and ammonia dissolved in pools (as they surely would) and that as these pools evaporated, the compounds became more concentrated, and finally reacted together to form more complex organic molecules. Another principal suggestion is that these relatively simple compounds were first held ('adsorbed') on the surface of clays, which then acted as catalysts, and caused them to react together.

A third notion is that organic life – carbon-based life – was preceded by 'living' clays, based on silicon, as described by Professor Cairns-Smith; that some of these clays gathered organic molecules around them, which increased their chemical versatility, and that the organic components eventually abandoned the silicon-based templates that had given rise to them.

An intriguing, though not widely accepted notion is that the first large organic molecules arose in very special circumstances, such as the hot springs (hydrothermal vents) that well up from volcanoes at the bottom of the sea.

Finally, there is the suggestion that organic molecules arose elsewhere in the Universe, perhaps on dust particles in space, and were first carried to Earth on meteorites. This notion is intriguing, and acknowledges the fact that carbon is actually *more* common in much of the rest of the Universe than it is on Earth. We know, too, that meteorites often do contain complex organic molecules. But of course, the 'life from space' hypothesis does not answer the question of how the very first organic molecules arose; and it seems at least as likely that they arose on primitive Earth as elsewhere. The idea is still discussed at scientific conferences, but most biologists feel that as ideas go, it is a bit of a dead end.

But although organic molecules are the essential components of modern living things, they cannot in isolation be considered living. How then, did these early, isolated molecules, take on the trappings of life?

The rise of life, and the idea of the cell

As we have noted, the 'things' that we now recognize as living are *organisms*. They contain complex organic molecules; but they are not just bundles of organic molecules. All the different complex organic molecules *work together* to form the entire organism.

Modern living things are organized in one of three main ways: as viruses, prokaryotes, or eukaryotes (see p. 14). Among the eukaryotes, some organisms consist only of single cells, including some fungi, such as yeast; and the 'protista' - single celled 'plants' and 'animals'. But some eukaryotes, including all the animals and plants with which we are most familiar, consist of many, often billions, of cells.

Running through all these organisms is a common theme. Nucleic acids, and in particular the kind of nucleic acid known as DNA (deoxyribonucleic acid), form the *genes* of the organism. DNA does not exactly 'make' proteins, but it provides a code which determines what form proteins should take. In its 'manufacture' of proteins, DNA is assisted by a subsidiary nucleic acid, called ribonucleic acid or RNA; and in some viruses RNA completely replaces DNA, and itself acts as the genes.

The 'creation' of proteins by DNA is significant because many proteins act as *enzymes*. These are biological catalysts, and they control the crucial steps in the workings of the cell: that is, in the cell's *metabolism*.

Thus proteins (in the form of enzymes) largely determine the second-by-second activity of the cell (the cell's metabolism); and also account for much of the structure of each cell. DNA determines what kind of proteins are made. Hence DNA – the genes – determines what kind of creature any particular organism actually is, and how it functions. DNA may be seen as the cell's administrators; proteins as its workforce; and RNA as the executives between those two crucial classes of molecule.

In modern organisms of all kinds, DNA and proteins work very closely together. DNA effectively 'makes' the proteins; or at least, it provides the code which, after due translation, determines that each protein is formed according to a prescribed pattern. On the other hand, DNA cannot replicate (reproduce itself) without the aid of enzymes, which are proteins. Thus without DNA, proteins would not exist in modern organisms; but without proteins, DNA could not exist. It is not fanciful, indeed, to regard modern life as a dialogue, between DNA and proteins.

DNA is a macromolecule; that is, a molecule of indefinite size (it can be very large indeed) which is compounded of lots of smaller molecules. Each of these smaller molecules is called a *nucleotide*. Proteins are also macromolecules; in this case compounded from molecular units called *amino acids*.

We have already looked at ways in which complicated organic

16

molecules could have come into being when the Earth was young; and these complicated organic molecules could well have included nucleotides and amino acids. If we are content to think generally, then, we can see how the key molecules of life might have arisen.

Yet there is a problem. The first amino acids and the first nucleotides – or the molecules from which modern amino acids and nucleotides later evolved – must originally have arisen quite separately. But although amino acids might well have become strung together to form primitive proteins, it is difficult to see how those proteins could have had any kind of coherent structure, or multiplied themselves efficiently, without help from DNA. Similarly, different nucleotides could have strung together to form primitive nucleic acid; but it is difficult, now, to see how that primitive nucleic acid could have replicated itself efficiently, without the help of some primitive protein.

In short, an essential step in the evolution of all the life forms that now exist on Earth, was the *co-operation* of two quite different kinds of molecule: nucleic acids based on nucleotides, and proteins consisting of amino acids. Only when this co-operation got underway could modern life progress. Biologists call such co-operation *symbiosis*; two different kinds of organisms (or molecules) co-operating to the extent that each relies upon the other for survival.

Charles Darwin, in his theory of evolution by means of natural selection, stressed the role of competition. Natural selection works *because* organisms are thrown into competition, and the 'fittest' are the ones that survive. Biologists who have come after Darwin have tended to focus upon the role of competition in driving evolution. Not only biologists: politicians, too, have sometimes liked to argue that political systems based upon competition are in some sense 'natural', because competition is such an essential part of nature. Alfred, Lord Tennyson, gave resonance to this notion, when he spoke of 'Nature, red in tooth and claw'.

But Darwin was a much better biologist and thinker than most of those who have come after him. He also stressed that organisms often survive best not by competing with each other, but through co-operation. And indeed when we ponder the relationship between DNA and proteins, we see that co-operation has been at least as significant a force in evolution as has competition, and at the most fundamental level. Indeed we should see life not as an ineluctable fight to the death, but as a delicate balance between competition and co-operation.

It now seems likely that the eukaryotic cell itself – the basic structure from which are constructed the bodies of all animals, plants, and fungi – originally evolved as a symbiotic association between various bacteria; that is, between various different prokaryotes. The thread of co-operation runs through all of life. The evolution of the eukaryotic cell is discussed more fully overleaf.

This, then, is life: innately complex organisms, cocooned upon a much-favoured planet, competing with each other for survival, but also co-operating with each other to their mutual benefit. The

The making of the eukaryotic cell

Co-operation takes many forms throughout nature. Many gregarious species obviously co-operate with others of their own kind, to the benefit of all: lionesses co-operate in hunting, and in bringing up the cubs; meerkats stand guard over the colony. There are also thousands upon thousands of close and interdependent relationships between species, and these are the kind that are called *symbiotic*.

But one of the most important examples of co-operation in nature is far from obvious. Many biologists, and in particular the American Llyn Margulis, suggest that the first eukaryotic cell was formed through a symbiotic association of different prokaryotes; by a coming-together of bacteria.

There are various kinds of evidence for this. In particular, eukaryotic cells carry within the cytoplasm, which surrounds the nucleus, various 'organelles'. These are discrete structures that carry out specific functions. Thus, virtually all eukaryotic cells contain mitochondria, which contain many of the enzymes that carry out respiration; the mitochondria are commonly known as the 'power-houses of the cell'. Plant cells carry chloroplasts: organelles containing the pigment chlorophyll, which they employ to entrap sunlight. In various ways, mitochondria and chloroplasts resemble bacteria. In particular, and most intriguingly, they both contain small amounts of DNA – and some characteristics are inherited via these mitochondrial or chloroplastic genes. It is odd that organelles, in the cytoplasm, should contain DNA at all *unless* they originated as complete organisms. Furthermore, in some respects mitochondrial DNA resembles that of bacteria.

In addition, many eukaryotic cells have whip-like flagellae or cilia at their surface, which may move the cell along, or serve to waft material past the cell surface. Thus, mammals have cilia in their lungs which remove dust, and mussels have cilia on their gills which shift a stream of mucilage which in turn traps nutritious detritus. Some bacteria, such as the one that causes syphilis, have flagellae; and it seems likely that flagellae and cilia *originated* as parasitic bacteria which, instead of causing disease, remained and became incorporated into the cell as the relationship was mutually beneficial.

If this hypothesis is correct – as more and more biologists believe it to be – then it illustrates that symbiosis, co-operation between organisms, is a crucial evolutionary step in nature; for without it, organisms as complex as plants, animals, and fungi simply would not exist.

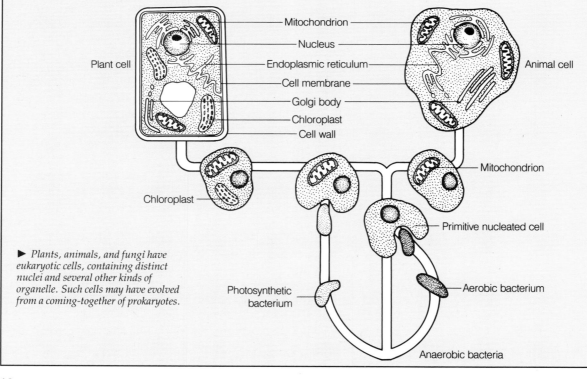

Plant cell

Mitochondrion
Nucleus
Endoplasmic reticulum
Cell membrane
Golgi body
Chloroplast
Cell wall

Animal cell

Chloroplast

Mitochondrion

Primitive nucleated cell

Aerobic bacterium

Photosynthetic bacterium

Anaerobic bacteria

▶ *Plants, animals, and fungi have eukaryotic cells, containing distinct nuclei and several other kinds of organelle. Such cells may have evolved from a coming-together of prokaryotes.*

interactions between them, and between them and their surroundings, are the subject of ecology.

We have stressed that the complex reactions of organic chemistry that are the processes and machinery of life, must take place in a watery medium. We could indeed extend our theme of symbiosis, to suggest that modern life at its most fundamental is a co-operation between organic, carbon-based compounds, and water. The range and significance of water in the processes of life is the subject of the next chapter.

2 *The waters of the world*

The magical properties of water

If 'life' is defined broadly; perhaps, as we suggested earlier, as any kind of chemical system that can evolve adaptively, then it is possible to envisage life of a kind that does not involve water. By the same criterion, it is even possible to envisage life without carbon. But in practice, modern Earthly life is a protégé of water, as much as it is of carbon. And just as carbon has very special properties, which make it so suited to the task – so too does water.

Most obviously, water is an excellent building material: eminently flexible, but not very compressible. Within living cells, it is held in a three-dimensional network of proteins to form the sponge-like cytoplasm, and this in turn is enclosed in a flexible protein-fat membrane. Even without the additional support of a skeleton, such building blocks can be moulded into an infinite variety of tough and versatile forms. Many highly successful modern animals, from earthworms to octopuses, rely almost entirely upon hydrostatic pressure to create their body forms. Indeed, animals with shells (the first kind of skeletons) appeared only 530 million years ago at the beginning of the period known as the Cambrian – fairly late, compared with the age of life itself. Plant cells typically contain a sac of water within the cytoplasm (a vacuole) which is held in place by the cellulose cell wall, and the whole water-filled cell provides an excellent building block. Herbaceous plants may be almost devoid of additional thickening.

Water has extraordinary physico-chemical properties, too, that serve life's purposes so well. To begin with, it is an outstanding solvent: more different kinds of atom, ion (charged atom) and molecule can be dissolved in water than in any other naturally occurring material. Earthly life is inveterately complex; of all natural materials, only water could bring together such a rich variety of elements, so conveniently, as this life requires.

As a bonus, water has particular electro-chemical qualities that make it an ideal reagent in many of life's metabolic processes. It is compounded, broadly speaking, of two atoms of hydrogen attached to one of oxygen: H_2O. In practice, however, the molecules tend to break apart: one of the Hs, carrying a positive electric charge,

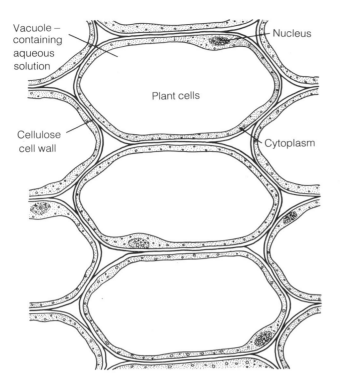

Vacuole – containing aqueous solution

Nucleus

Plant cells

Cellulose cell wall

Cytoplasm

becomes detached from the remaining OH, which is then known as a hydroxyl radical, and carries a negative electric charge. Living things constantly transform one kind of molecule into another, during digestion, assimilation, respiration and repair; and many of these transformations are effected by attaching hydroxyl radicals, ultimately derived from water, to the particular organic material that is being attended to. In addition, photosynthesis (see p. 93) involves the controlled splitting of water.

Water has a greater capacity to absorb heat than any other natural material except for ammonia. It takes one calorie* of heat energy to raise one gram of water through one degree Celsius. For most materials – like iron and stone – it takes only a tiny fraction of a calorie to raise the temperature by one degree C.

At least 80 per cent of a typical living cell consists of water; and its ability to absorb heat without raising temperature protects all the delicate organic molecules that it contains. On the global scale, the oceans that cover almost 70 per cent of the world prevent it from

▼ ▲ ▼ *Water is an excellent building material. Many creatures, even large herbaceous plants and octopuses (above), are held up by hydrostatic pressure. Plant cells contain a vacuole of water held in place by a cellulose cell wall (above left), and leaves wilt if the water evaporates faster than it can be replaced (below).*

* 'calorie', spelt with a small 'c', is defined as the amount of heat required to raise one gram of water through one degree Celsius. For the purposes of nutritional science, however, the calorie is too small a unit to be useful. So nutritionists employ a bigger unit – the kilocalorie, which is 1000 calories. Kilocalorie should be written 'kcal' or 'Calorie', with a big C. But people who write about food (nutritionists and non-nutritionists alike) tend to write 'calorie' when they mean 'Calorie'. So bear in mind that when food writers say that an adult man needs around 2000 calories per day, they really mean that he needs 2000 kcals, or Calories: that is, two million calories!

becoming intolerably hot by day, and freezing cold by night. We get some glimpse of what conditions would be like without water by observing modern deserts, which commonly fluctuate from +45°C to around zero, in the course of twelve hours!

We take for granted, too, the fact that water is liquid at normal Earthly temperatures and pressures: pure water freezes only at 0°C, and boils at 100°C (which is how those quantities are defined). Yet chemists would guess, were it not for the evidence of their own senses, that a material containing two hydrogen atoms and one of oxygen in each of its molecules should be a gas at Earthly temperatures – like carbon dioxide, with one atom of carbon and two of oxygen, or nitrogen dioxide, with two atoms of oxygen and one of nitrogen. How fortunate that it should be liquid, at precisely the kinds of temperatures at which complex organic molecules, such as proteins and DNA, are able to persist and yet be chemically active!

As an encore, water miraculously expands, just on the point of freezing; in fact, as its temperature falls from 4°C to 0°C it expands by 10 per cent. Thus, ice floats. If water continued to contract as it froze, like most other materials, ice would sink. Having sunk, it would then be insulated from the Sun's warming rays by the liquid water above it. Such hypothetical 'heavy' ice would therefore take much longer to melt than real ice does. Thus, many more of the world's waterways would be permanently frozen.

The influence of water upon living things does not end here. The most obvious quality of water, after all, is that it is liquid. Because it is liquid – by definition – it can move. The movements of water affect both the surface structure of the land (its *topography*); the overall climate; and the day-to-day weather. As the lives of animals and plants are clearly influenced profoundly by topography, climate, and weather, the movements of water are clearly of profound ecological significance.

Why, first of all, does water move?

The movements of water

Water moves for five main reasons.

The waters of the world's oceans are tugged outwards by the gravitational pull of the moon: a hill of water forms in the middle of the ocean when the moon is appropriately positioned above it. As the moon passes, the waters relax and spread. Hence the tides. The orbit of the moon is such that its gravitational pull is great one week, and smaller the next, then great again. Thus big 'spring' tides, when the sea rises and ebbs most markedly, alternate with neap tides, when the rise and fall is much less pronounced.

Secondly, the gyration of the Earth on its own axis tends to cause the oceans to spin.

Heat from the Sun also causes the waters of the world to move. This heating is uneven: more by day than at night, and more at the equator than at the poles. In addition, the oceans are in contact with the land. The rocks of the land have a smaller specific heat than water

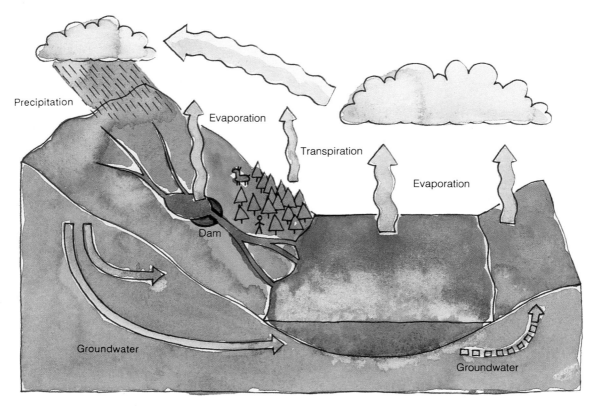

does, which means that any given amount of heat that falls upon them will raise their temperature more than that of water. So land tends to be hotter than the surrounding sea by day, and cooler by night; and in general, hotter in summer and cooler in winter. Heat flows from regions of high temperature to regions of low temperature, so that heat is either flowing from the continents to the oceans, or from the oceans into the continents. Water that is above 4°C expands as it is warmed, like any other material. As it expands, so its density is reduced. Like all liquids, water that is less dense tends to float upon water that is more dense. Thus, as the world's oceans are unevenly heated and cooled – more in some places than others, more at some times than others – so it tends to shift. Because these movements are superimposed on movements caused for other reasons (such as the moon's gravity), and because the continents are in the way, these movements are very complicated indeed.

Each year, from the surface of the ocean alone, 300 000 cubic kilometres of water are lifted into the atmosphere by evaporation. Most of this falls back into the sea, as rain or snow, but 100 000 cubic kilometres of it falls on land. There, it percolates down through the rocks into aquifers (underground water reserves), or flows as surface rivers back into the sea, propelled by the Earth's gravity. This mass movement of water, ultimately powered by the Sun, is known as the 'hydrological cycle' (see above).

◄ The hydrological cycle.

Finally, the world's waters differ greatly in the amount of minerals dissolved in them. Oceans have been picking up soluble materials (notably sodium chloride – salt) from the land, for millions of years. Water that has evaporated and condensed in the upper atmosphere, and then falls as rain, should be pure, except for a little carbon dioxide picked up on the way down (although in fact, these days, it tends also to contain a whole catalogue of pollutants). Water that is salty is more dense than water that is fresh. So as less salty water comes into contact with more salty water, the two water masses shift (the fresher mass tending to float on the saltier mass). Eventually things even themselves out, as the salt spreads itself evenly between them; but this takes time, and the movements occur first. On page 42 we will see the signficance of this in helping to generate one of the world's most important oceanographic and climatological phenomena, the warm southern ocean current known as El Niño.

All these movements affect living things in many different ways. A direct and obvious example is that many marine animals are dispersed around the globe by ocean currents. But the indirect effects may be more profound. Thus, the movement of water is the chief sculptor of landscape – after Earth movements have put the land in place – and it is one of the chief determinants of overall climate and of day-to-day weather. Landscape (topography), climate and weather profoundly affect the lives of living things. Indeed they largely shape their evolution and their behaviour, and determine which creatures can live where.

Water – the sculptor of landscape

The continents are where they are and have the mountains they have because of the action of the Earth itself; the eruption of volcanoes and the shifting of plates of the Earth's crust (plate tectonics). But once the land is thus rough-hewn it is shaped by other forces. The Sun itself is destructive, heating the rocks by day so they expand, abandoning them to cool at night. But the greater destroyers are water and air, and the materials they carry. Air, especially when laden with grit, should not be under-estimated. The softer stones of ancient cathedrals show how deeply it may cut; and the rock from which the Sphinx was carved may have first been roughly shaped by desert winds, for the general outline follows precisely the aerodynamic flow of air over obstacles, swirling away at the end to leave a mass for the head; and there are many rocks of similar form elsewhere.

But water is the greatest shaper of landscape. It acts in a wide variety of ways, exerting each of its chemical and physical qualities. Some rocks, such as limestone, it may simply dissolve away – the more so if the water is acidic. Fissures in rocks are widened as water freezes and expands within them, exerting tremendous forces as it does so. This initial fragmentation is one of the essential processes that leads to the conversion of sterile rock into fertile soil.

Water may break up rocks, too, simply by pounding against them, as waves do. Or it may abrade; picking up silt, sand, or even rocks

▼ Water is a sculptor of landscape; it gives, and it takes away. Here, the sea erodes the land. But further up the coast, the land that is removed may be re-deposited.

and grinding them against the rock beneath. If the water is frozen, and shifts as a glacier, then it can carry entire boulders, as big a house or as big as a hotel, hundreds of miles, and gouge great scars.

But water is a creator of land as well. Silt, sand, and rocks are heavier than water. They are carried by it only if the water has enough energy to lift and shift them – that is, if there is a great enough mass of water, moving sufficiently quickly. If the water is slowed, then these heavy materials come out of suspension, with the light material (silt) being carried further than the heavier. Water may be slowed when rivers meet the sea, or by bends in rivers, or – depending on the local topography – by friction as currents move along a shore. Then the load is dumped, to form beds of silt in river-deltas, or river banks, or dunes or offshore spits. Often, as the new feature is formed, so it creates new currents, which begin the destruction all over again. Some water-formed features therefore, including the sand-bars that may form parallel with a shore-line with a salt-marsh behind, are constantly formed, destroyed and re-formed.

Nature is not fixed. Sometimes human intervention is not a question of changing the environment, but of seeking to prevent its change. In particular, people invest a huge amount of money and effort in creating coastal towns, for the coasts are good places to live. But coastal towns, once built, must often invest millions of pounds in groins and sea-walls and often in wholesale beach replacement, simply to keep the coastline more constant than unaided nature would allow. Often the battle is lost, and many a seaside town has disappeared beneath the waves (while others, such as the once-thriving port of Rye in Kent, have been left high and dry as the sea retreated). Such engineering may enrich wildlife, by creating more crannies for coastal creatures to live. But it may also be destructive. For example, beaches are sometimes naturally shifted from one area and deposited up the coast. If the sea is prevented from carrying some particular beach away, then another place, up the coast, may be robbed of its fresh supplies, and simply disappear. Thus well-protected holiday beaches sometimes take a toll from less favoured areas.

Tides are destroyers and creators of coastlines – nibbling here, building there; it is odd to think that our coasts are shaped largely by the gravitational energy of the moon. In addition, the constant incursion and retreat of the sea creates and maintains vast zones that are neither wholly marine, nor wholly freshwater, nor wholly land-based; *intertidal zones* that vary in width, between spring tides and neap, and which could not persist, or would not exist at all, were it not for the tides.

Such zones take various forms, depending on the shape of the continental edge, its chemical and physical properties, and its age. Sometimes the intertidal zone is a beach; sometimes part of the wall of a cliff. Salt marshes, too, would not exist were it not for the tides. If the sea did not constantly encroach, the salt within them would be washed out by rain.

▶ *Salt marsh is a rich habitat with a host of specialist species – such as glasswort,* Salicornia, *shown here. Were it not for the tides, it is hard to see how salt marsh could exist at all.*

The mud-flats at the mouths of rivers, dumped as silt when the river meets the sea, tend to accumulate minerals and organic materials. Thus they support vast numbers of creatures, and in particular provide the principal feeding ground for an extraordinary variety of specialist birds, the waders. A stretch of land or water, with the air above it and all the creatures therein – all the living and non-living things that live and interact together – is termed an *ecosystem*. It is hard to envisage how an ecosystem such as that of the mud-flats could be maintained, were it not for the tides.

Water – the creator of climate and weather

Water can contain vast amounts of heat energy. This energy heats the air above, which rises. This basic movement is complicated by a number of factors. The water itself shifts, and thus carries vast masses of heat with it; the land also heats the air, and is often at a different temperature from the water; the heating is greater at the equator than the poles, and greater at night than by day; the air is diverted by mountains, and funneled through passes; and the air itself is heated by the Sun. Air, too, is powerful stuff *en masse*, and as it moves, so it in turn agitates the water beneath: storms are initiated by winds (although the winds themselves may derive their energy from the heat of the sea).

Thus the warming and the movement of water help to give rise to climate, which is immensely complex; and the day-to-day fluctuations which are known as 'weather' are extraordinarily difficult to predict in detail, more than a few hours ahead.

Why life is 'patchy'

Water is the perfect, and 'natural' medium of life; obliging in its chemistry; providing constancy of temperature, and support. Naïvely, then, we might suppose that all bodies of water throughout the world would be evenly packed with living things. Yet this is not so. Some areas seem particularly favoured, while others are a virtual desert. Conservationists, and all who seek to exploit the riches of the world's waters, must take particular care of the favoured areas, for they in many cases are the fountain-head of all the rest. But why are some areas so favoured?

You would expect the tropics to support more life than the high latitudes, simply because they receive more solar energy. And so, in general, they do. But other factors play a part – and because of them, we often find regions that are extremely well endowed with living things, even in high latitudes, and often in areas where common sense suggests that life might be most difficult.

Thus, intertidal zones and estuaries in particular tend to be extremely rich, even though they are battered by waves or scarified by rapid waters, are sometimes dry and sometimes wet, and pose all kinds of chemical problems as the salinity may swing from fresh to super-saturated. Some ponds and lakes, too, are richer in life than might be expected; and sometimes, indeed – when polluted – become too rich for their own good (this is known as *eutrophication*; see p.117). Other regions that on the face of it may seem rather benign – including open reaches of tropical ocean – may, for long periods, be almost devoid of life. What are the factors, then, that cause such unexpected patchiness?

In water, and to a lesser extent in forests, the first of these factors is light. Most life on Earth depends, in the end, upon photosynthesis; and that can take place only in the light. Sunlight penetrates down into clear ocean at most about 200 metres. But if the surface is full of light-scattering particles (including the floating animals and plants of the plankton) then the 'photic zone' (the region of light penetration) is far less than this.

The second complicating factor is availability of nutrients, which in bodies of water, depends largely on the degree of mixing. So far we have stressed the general mobility of water, but in the oceans much of this motion is horizontal, rather than vertical. Indeed, because the oceans are heated from above, and warm water expands and is less dense than cold water, the warm water tends simply to sit on the surface, unless otherwise disturbed. Typically, each spring, the planktonic plants and animals in the surface photic zone do very well, feeding on nutrients at the surface. But when they die they sink to the bottom, taking the nutrients with them. Through most of the summer, then, the plankton in open ocean is gradually depleted. Fertility is not restored until the winter storms stir the waters again. Near continents, there is more mixing, as currents and tides strike the land. Hence there is more plankton. Major fisheries – those for cod of the North Sea, the anchoveta of the Eastern Pacific – occur in areas

▶ *Fisheries are sustained by the plankton which depends upon a constant re-cycling of nutrients stirred up from the sea bottom. Thus the world's biggest fisheries occur close to continents.*

▼ *Unlike most animals, mussels do not expend energy searching for food. They wait for the tides and currents to bring it to them. Thus the currents serve them as an energy subsidy.*

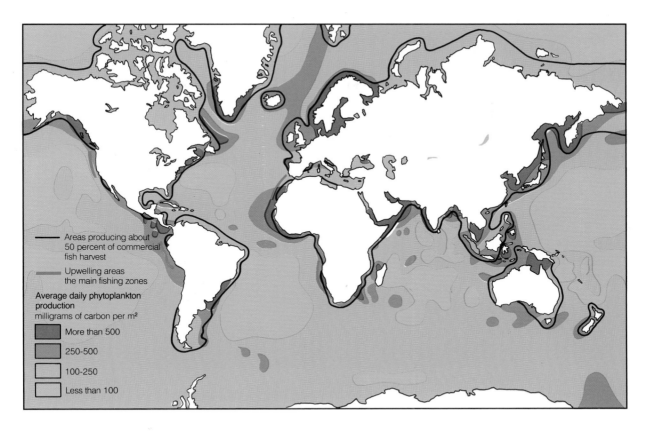

Areas producing about
50 percent of commercial
fish harvest

Upwelling areas
the main fishing zones

Average daily phytoplankton
production
milligrams of carbon per m²

More than 500

250-500

100-250

Less than 100

where, for various reasons, nutrients are stirred up from the bottom
and the plankton can thrive.

The third major cause of patchiness is energy subsidy. Some
favoured ecosystems receive energy additional to that of sunlight,
both in the form of organic molecules (or other nutrients) brought
from elsewhere, and sometimes in other forms, too. This is where
estuaries score very highly. Obviously, the same river flow that fills
them with silt also brings in minerals and organic materials from
elsewhere. Deposit feeders, such as many gastropods which ingest
mud, digest out the organic molecules, and excrete the rest, benefit
from this. Filter-feeding animals such as mussels benefit from the
movement of the tides, which brings their food to them. They do not
waste energy, as most animals are obliged to do, in pursuit of their
prey. The movement of tides is an energy subsidy.

Such phenomena as this are natural. Estuaries are generally
enriched. Ponds and lakes, too, benefit from their surroundings, as
leaves and insects perish within them, and some nutrients run in
from the surrounding land.

Various modern human activities ensure, however, that many
ecosystems are assaulted with more nutrient than they can properly
deal with. Worldwide, indeed, surplus nutrient is at least as
dangerously polluting as pure toxin. In ponds and lakes, surplus
phosphorus in particular, but also nitrogen, commonly from fields

29

but also from domestic sewage, may cause tremendous overgrowth of cyanobacteria or of green algae, far more than the resident animals can possibly consume. These have a rapid turnover, and as they die the bacteria responsible for their decay use up all the oxygen in the pond, thus destroying all animal life. Such overgrowths are known as *blooms* (see p. 117).

Many animals that do not inhabit the especially favoured areas all the time, none the less make use of them in the course of their lives. For example, migrating wading birds may stop off at particular estuaries *en route* from winter feeding grounds to summer breeding grounds; and we will meet more examples (eg of fish breeding in mangroves or coral reefs) in the next chapter. But until we study the life cycles of animals in fine detail, we cannot know precisely which creatures depend upon what; and many a creature has been wiped out not because its entire habitat has disappeared, but simply because some favoured site has been obliterated.

Often the places that accumulate nutrients, and so are naturally favoured, are also the most vulnerable. They are prone to accumulate more nutrient than is good for them; and, of course, they tend also to accumulate toxins. This latter tendency is exacerbated because many organisms have a natural ability to extract and concentrate minerals from their surroundings, even if these are present only in very low concentrations. Thus tunicates (creatures distantly related to vertebrates, and commonly known as 'sea-squirts') naturally accumulate vanadium or titanium which are present in sea water only in a few parts per billion.

30

◄ ► *Aquatic algae, like all plants, need nutrients. But when too many nutrients flow into bodies of water, in sewage (right), fertilizers, or perhaps in detergents, algae may grow excessively to form a 'bloom' (left). The over-enrichment of bodies of water by nutritious pollutants is called 'eutrophication'.*

Tunicates accumulate those rare metals for physiological purposes. Virtually all creatures, though, are capable of scavenging rare materials from the environment and accumulating them effectively by 'mistake'. Thus all animals living in the wild used to accumulate DDT in their fat stores, in the days when DDT was a common pesticide. All of us pick up and accumulate lead, distributed around the environment from car exhausts. To return to our marine theme, fish living in Minamata Bay in Japan in the 1950s accumulated mercury, discharged from a nearby chemical plant. Mercury in metallic form is not particularly dangerous, but many of the organic compounds that can be made from it are extremely toxic. Some of the mercury was already in organic form at the time of discharge, and the mercury compounds underwent further 'biotransformation' inside the fish, which rendered them extremely dangerous. These mercury compounds had ghastly effects on the human nervous system, causing convulsions, blindness, and death. By the 1980s 300 people had died from eating the poisoned fish, and at least 1500 more were disabled. Poisoning of waterways is hideous enough; but the ability of many organisms to concentrate toxins makes it even more so.

Because the world's streams and rivers provide such convenient sewers, and because the seas and lakes seem all-accommodating, human beings have, over the years, abused them notoriously. This continues, but at least now efforts are also being made to prevent further pollution and even repair some of the damage. How far is this possible?

31

The tragic collapse of Lake Victoria

For thousands of years, deliberately and unconsciously, human beings have been carrying animals and plants – and microbes – into new territories. Sometimes the interlopers have simply died out. Sometimes the newcomers have settled in without doing the original inhabitants any great harm – and so it is that the rabbit and the fallow deer have become accepted additions to Britain's fauna.

But sometimes – usually – introduced species have had disastrous effects on the native flora and fauna, and sometimes on the human inhabitants. Thus dingoes, taken into Australia by Aboriginal people, probably about 3000 years ago, out-competed the thylacine (marsupial 'wolf'). Much later, the Europeans brought cats to Australia (which probably jumped ship in the seventeenth century), and then foxes and rabbits (for sport) in the nineteenth century – and these have out-competed and preyed upon a whole host of marsupials. Starlings and a wide range of plants from the Mediterranean (some introduced for ornament, some escaping as weeds from agricultural crops) have made deep inroads into the native animals and plants of North America. The Mediterranean, in its turn, has been overrun not only by tourists, but also by cacti from North America. Worldwide, indeed, 'exotic' species introduced by human beings have become a major cause of extinctions: probably second in importance only to loss of habitat.

Indeed the most dramatic example of mass extinction known to biology has been caused by the introduction of an alien species. It has occurred in Lake Victoria in East Africa.

Lake Victoria is vast – the size of Switzerland. Until recently, it contained a wide group of fish of the kind known generally as haplochromines: that is, they were once classed in the single genus (see p. 124) *Haplochromis* within the family Cichlidae, though they are now generally placed in several different genera. Individually, haplochromines are not spectacular in appearance: most are only a few centimetres long, and silvery, when they are not in their breeding colours.

Yet the haplochromines were one of the biological wonders of the world: for within that single lake, they had diversified into 300 different species. Each of the 300 occupied a slightly different ecological niche: some living among the rocks inshore; some in the depths; all (as many cichlids do) holding their developing young in their mouths for protection ('mouth breeders'); and some highly specialized types earning a living by sucking the young ones

out of the mouths of brooding mothers. No one knows how they got to be so diverse, though Dr Humphry Greenwood, of London's Natural History Museum, has suggested that the water in Lake Victoria has at times been lower than now, with the fringes separated into pools. Populations evolved separately in each of these pools, and only when the water rose again did they re-convene in the same lake.

The haplochromines were also extremely numerous; indeed, they were by far the principal biomass of Lake Victoria. The local people caught them in vast numbers for food, eating them fresh or drying them in the sun, and until recently they made up 80 per cent or more of the total catch.

Then in the early 1960s Nile perch, of the genus *Lates*, were introduced into Lake Victoria. *Lates* can be huge – some up to two metres long, and weighing 150 kilograms. They occur in many other lakes and rivers in Africa, but have never lived in

the present Victoria (though they did live in a prehistoric lake on the same site).

The Nile perch flourished. By the 1980s, indeed, they commonly accounted for 70 per cent of the Lake Victoria catch. They flourished by eating the haplochromines – which now make up only about one per cent of the catch. More to the point, extensive studies by Dutch biologists suggest that as many as 200 out of the 300 original haplochromine species are now extinct.

Economically, the Nile perch has been a mixed blessing. For the present, it has brought income: the Kenyans now export Nile perch, even to Europe. But it is clear that the present ecology of the lake cannot be sustained. Nobody knows what the Nile perch will feed on when the haplochromines have been further reduced in biomass. Already there are indications that some are feeding on their own young. Furthermore, the lake ecology is now upset, and the waters are becoming eutrophic, with a build up of algae and dead plankton. Large areas of the lake are now lacking in oxygen (anoxic).

Conceivably the entire lake could die within a few years. It would not be the first to do so, though it would be the biggest. Biologists can only watch and wait.

For this reason alone, the introduction of *Lates* to Lake Victoria may soon be seen as one of the great biological disasters of the twentieth century. Scientists interested in biological diversity, and the evolutionary reasons for it, already see it in that light.

Other African lakes, such as Malawi and Tanganyika, face other threats. Oil has been found under both lakes, for example, and at least one company would like to drill for it. The protection of what is left must be regarded as a priority.

▼ *Nile perch: a fine food fish, but responsible for one of the greatest biological disasters of the twentieth century. It seems to have caused the extinction of up to 200 species of haplochromine fish in Lake Victoria.*

The desalination of Australia

Farmers and governments everywhere would like to make more productive use of the vast areas of the world that are arid or semi-arid; land loosely classed as 'desert'. The rewards from deserts are never going to compare with those from places with abundant water: and most of the world's deserts are best left to the highly specialized animals and plants that are adapted to live in them. But some exploitation is possible and necessary, for many hundreds of millions of people live in such areas, and have nowhere else to go. The desert has ways of showing, though, that exploitation is even more difficult than is at first apparent.

One of the key problems of desert exploitation worldwide, is that of salinity: an increase in salt at or near the surface. In the Third World, as in much of the Punjab and Sudan, such salination is of critical significance. In affluent Australia it is at least highly inconvenient, and has driven many a farmer into bankruptcy.

About 70 per cent of Australia is arid. Of this, almost half is used as rangelands, for raising cattle and sheep; a total area somewhat greater than India. Of this, 15 to 20 per cent is already severely degraded. Salination is only one of the problems, but it is a serious one.

Australia is extremely flat, for a land-mass of such size. Whatever falls on it, stays where it falls. In Australia's dry soils there are millions of tonnes of salt blown in from the sea over many millions of years.

This in-coming salt of course is deposited on the surface; but over time, washed down by the (albeit occasional) rains, it tends to accumulate in the ground water. Soon the ground water becomes brackish. This contaminated water eventually re-surfaces as a spring, and effectively poisons the plants all around.

For millions of years, however, the native plants of Australia have prevented the accumulated salt from sinking into the ground water. Deep rooted trees, notably the eucalyptus and acacias (known in Australia as 'wattles') draw water up from the depths of the soil. So too, to a lesser extent, do the herbaceous plants of the bush (9000 species in Western Australia alone), which stood belly-high to a horse when the first Europeans arrived in the late eighteenth century. And the growth of these herbs was in turn encouraged by the Aboriginal people – paradoxically, by setting fire to them, and getting rid of whatever was rank.

The Western sheep and cattle rangers cut down the big trees and put a stop to the burning of bush.

▲ *Australian agriculture has swept aside many plants of the bush which used to prevent salt from sinking into the groundwater. Now much of Australia's desert is salinated.*

They sowed exotic grasses for their animals, built up the herds during the good years, when it rained, but then (at least in some cases) found that the grass was overgrazed in the drought years, when it failed to grow. The meagre vegetation that remained could not prevent such water as there was from sinking into the aquifers beneath, and taking its load of salt with it. Where the land is particularly low-lying, this deep-lying, brackish water comes to the surface. These are the areas now ruined by salt.

However, Australian conservationists feel that in this, as in many aspects of environmental protection, the onus is on them to show the rest of the world how to cope with the problems. Australia, after all, shares the problems of other tropical countries – possessing both tropical forest and semi-arid bush. But unlike most countries of the tropics, it is not itself poor. It has the money and the science to cope.

The basic solution is first to restrict the flow of water percolating down into the aquifers by planting trees, preferably native species, on the higher ground. With tree-cover restored and thus degradation reversed, attention can turn to the lower ground, already over-salinated. One solution here is to plant salt-tolerant species. Notably, these include 'salt-bushes' of the genus *Atriplex*, a relative of spinach which excretes surplus salt through its leaves; or the remarkable tamarugo bush from Chile, related to the pea, which flourishes in areas where surface salt lies a metre thick. Such plants can be grown, harvested, and then disposed of (perhaps fed to cattle in some distant place); thus steadily reducing the salt load.

It will take decades to reclaim the degraded dry lands of Australia by such means. But much is restorable, and the effort is worthwhile.

Cleaning up the world's waterways

It is clear, first of all, that we can do a great deal to restore damaged waterways by using common sense and good engineering. For example, we can prevent factories that are obviously discharging their waste into rivers, from doing so; and there is virtually no kind of effluent nowadays that cannot be safely treated and disposed of, if we are prepared to spend the money. We can also, to some extent, reverse the damage caused by polluted waterways (see left).

Yet, as we have already seen, the waterways of the world are complicated. Lakes may take in water from many miles away that has percolated through the soil or through aquifers over decades. It seems highly likely, for example, that much of the nitrate now appearing in some ground water in Britain was actually first released almost fifty years ago, when grassland was ploughed to make way for cereal as the country strove to increase its home-grown food supply. In the sea, the complexities are compounded by tides and currents, bringing pollutants sometimes from thousands of miles away, and from the sheer number of possible sources of pollutant – sewage works, factories, and agricultural fields – which feed into the rivers which in turn run into the sea.

When a sea or large lake is polluted, then, it often is not possible to identify precisely the main sources of the pollution, or to predict the effects of any particular preventive measure. This is important, because clean-up measures are expensive, and funds are limited even in rich countries.

One partial solution, of the kind that is now almost universally adopted by serious conservationists, is to construct a mathematical, or computer 'model'. The computer is fed with all the relevant data, and ecological theory, and is then asked, 'if we did such and such, to this lake (or river, or bay, or whatever), what effect would this have on such and such a pollutant?'

Such models do not give perfect answers, but they do improve on common sense. An example of one in action is shown overleaf. In fact this is not an attempt to reverse pollution, but something more complicated: to create a plan that would enable Lake Ichkeul, in Tunisia, to be used for several different purposes simultaneously.

Similarly, the effects of altering the course or behaviour of a waterway are extremely complicated, can be very far-reaching, and take decades or centuries to unfold fully. The Aswan Dam, more than thirty years old, provides a mature example (p. 37). Again, computer models (incorporating knowledge gained from previous dams) can be used to some extent to predict the likely effects of future projects.

There are techniques, then, for minimizing damage to the Earth's waterways, even though we may choose to exploit them. But if we are to preserve the creatures that rely directly upon water as their medium, technique and science alone are not enough. Conservation cannot work unless we have the will to make it work. All conservation depends, in the end, on human attitude. What attitude is appropriate?

A plan for Lake Ichkeul

Once, North Africa was dotted with freshwater lakes. Most have been drained for agriculture, or otherwise polluted, but the biggest of them all remains: Lake Ichkeul, in Tunisia. It is one of the great foci of wildfowl from Europe and Asia. There are up to 400 000 birds present at a time: mostly greylag geese, wigeon, and coot, which spend the winter from November to February, and others which pass through *en route* to South Africa. Rarities come too, like the white-headed duck. Ichkeul, indeed, was only the second wetland, after the Florida Everglades, to be given World Heritage status. But Ichkeul is also acknowledged, now, to be among the world's most threatened wetlands.

Ichkeul is a complicated place. Only some of it – about 100 square kilometres – is fresh, for most of the year; and there are another 60 or so square kilometres of freshwater marsh. Because it is open to the sea, the lake becomes salty now and again, but the salt is flushed out by the rivers that feed it from the African mainland. It matters to the birds that the lake remains fresh, because they feed on the freshwater weeds such as *Potamogeton*, and rushes such as *Scirpus*. But the rivers that keep the lake fresh are required for agriculture; and there have long been plans to dam them.

Tunisia is not unaware of its environmental responsibilities. It has an Environment Protection Agency. And almost a decade ago its government asked scientists, including hydrologist Dr Ted Hollis from University College London, to predict the effects of damming the rivers upon the level and salinity of the lake, and hence upon the weeds and birds; and to suggest ways of minimizing the impact of damming.

Such prediction is immensely complicated. The easiest part is to work out what would happen to the waters of the lake itself; but for this, Dr Hollis had to pool information on rainfall, evaporation, the flow and salinity of the rivers, and the level, volume, and salinity of the lake, over as many years as possible (in practice thirty) and feed this into a computer.

The model shows that unless there is 'compensatory management', then damming the rivers will turn the lake saline, reduce its level, kill the present vegetation and effectively banish the birds. It also shows, however, that sluices judiciously placed, and a barrier across the lake to keep one area free of encroachment from the sea, could allow birds and agriculture to flourish in concert; an ideal indeed, for all the world.

At the time of writing, however – 1990 – it remains to be seen whether the necessary sluices and barrier will be built.

▼ *Ichkeul: one of the last and biggest of North Africa's great freshwater lakes. Plans are afoot to divide it equitably between farmers, who want its waters for irrigation, and conservationists, who seek to protect its wading birds.*

The Aswan Dam: a lesson in complexity

The economy and culture of Egypt since before the pharoahs was built upon the annual flooding of the Nile; upon the rains that fall in Ethiopia between August and November, and then flow north, through the Sahara, for 2400 km to the Mediterranean.

But the people of modern Egypt were not content with agriculture that was so seasonal, or was confined to an area that utilized only three per cent of their country. Beginning in 1902 they began a series of four dams, to regulate and to some extent re-direct the flow. The biggest was planned in the 1950s, for Aswan, in the south of Egypt, roughly 640 km from Nile delta. It was completed in 1970, at a cost of one billion dollars – a huge sum in those days. The Ancient Greeks had a word, 'hubris', to refer to any human action that over-reached itself – of the kind that the Greeks regarded as an offence against the gods. The Aswan Dam is modern hubris. Human science was not, and still is not, able to predict all its effects. We will look briefly at just four.

To begin with, dams create lakes behind them, and the one that was to build up behind Aswan, Lake Nasser, was supposed to be 500 km in length, with a surface of 5100 square km. The sandstone rocks along its western side were known to be porous, but the engineers predicted that the pores would be blocked by silt. The silt stayed at the bottom, however, and the bank remained porous. In the beginning, it lost about 15 billion cubic metres of water by seepage each year. It loses another 15 billion cubic metres from the surface by evaporation – a loss that is about twice what was anticipated, because of the unexpected effects of wind. These losses roughly equalled the total inflow into the Mediterranean before the dam was built.

The original flow of the river greatly affected the Mediterranean, too, and the marine life that was adapted to it. In particular, the annual flooding brought silt, which caused blooms of algae to develop which in turn supported a rich fishery. The fish harvested in the Eastern Mediterranean totalled 135 000 tonnes in 1964, the last year in which floodwater reached the sea unimpeded, with sardines accounting for 15 000 tonnes. By the early 1970s, the sardines had gone.

The irrigation canals that flow from the newly controlled river have also caused problems, because they carry the parasitic diseases malaria, bilharzia, and trachoma.

As one of many side issues, we may cite – appropriately enough – the Great Sphinx. This has stood for several thousand of years – or actually for millions, for it was carved *in situ* from natural rock. Now it is beginning to crumble, as the water from the newly created water table rises up within it and deposits salt beneath its surface. That such a monument should now require a damp course is a suitably bizarre commentary on human folly.

▼ *The Sphinx: probably carved from rock that had been roughly shaped by the wind, and therefore even more ancient than it seems. Now it is crumbling as water trapped by the Aswan Dam brings rising damp.*

A proper attitude to water

Various societies, at various times in history, have formally and 'officially' treated the world with reverence. Often these people believed that each of the components of nature – the forest, the rivers, the sky – was occupied and guarded by a jealous god, and they behaved as if that were the case.

Other societies, at other times, have simply regarded the elements of nature as commodities: sources of food, fuel, shelter, transport, and wealth. They have acted on the principle of 'out of sight, out of mind'.

This latter attitude is not unique to modern times, or to western people. But modern westerners (and those inspired by European culture) combine this way of looking at the world with advanced technology, enabling them to make huge changes, quickly.

The waterways of the world are suffering enormously from such carelessness. Since World War II, especially, technology-inspired governments have tended to regard the oceans virtually as a bottomless source of food for the world's billions, and of oil and minerals besides; and the world's great rivers are systematically 'harnessed' for hydro-electric power, and re-directed for irrigation and for transport.

At the same time humankind as a whole has treated the oceans, rivers, lakes and ponds as a dumping ground, for sewage, noxious chemicals, radioactive waste, and generalized junk. Often this has been done deliberately: factories and sewage works are *built* to discharge into seas and rivers, and 'honey barges' ferry cargoes of sewage out to sea. But spoliation has also been inadvertant, and many a Scandinavian lake has now been sterilized by rain made acid by the exhausts of motor cars and factories, hundreds of miles away.

But whether deliberate or merely careless, the outcome is the same. Whatever disappears beneath the surface can no longer be seen. A lake that has been poisoned looks just as blue beneath a northern sky.

Such an attitude to nature has always been hideous. Applied on the present global scale, it is downright dangerous. And it is particularly inappropriate when applied to the world's waters. The oceans are never going to be stupendously productive, relative to their size; high productivity is always localized, and when it is achieved it is always fragile. It would be foolish to suggest that a river should never be tapped for energy or for agriculture, but the world's politicians have not yet chosen to realize what enormous consequences such action has, or how long those consequences take to unfold, or – an essential consideration – that it is literally impossible to predict all that will ensue when a river is tapped.

Yet we do know, now, that the world's waters are the world's essential regulators: of global temperature, and of the chemistry of the atmosphere – which is nothing like so invariable as we might like to suppose. In addition, the world's waters are all interconnected; whatever is done to a river or even to some remote stretch of land will often, in time, be reflected in some alteration of the sea. Even the

▲ *Beautiful still, but empty. Because of acid rain, this Scandinavian lake is now too acid to support fish.*

polar ice contains pesticide, for example.

In short, the component of the world that we have treated most avariciously, and most insouciantly, is the very one that we should treat most deferentially. It is the one, above all, that is best left alone. We may not believe any more in gods inhabiting rivers and lakes and oceans. But it is in our own interests to act as if they were there.

These, then, are some of the ideas that attach to all the world's waterways. The greatest of all the world's waters, of course, are the oceans; and they are the subject of the next chapter.

3 *The oceans*

To the prejudiced eyes of land-bound humans the oceans seem like one continuous mass, homogeneous as outer space. To some extent they are, and some marine creatures treat the whole maritime world as their oyster. Some of the great whales, for instance, plunge from the surface to the depths as a matter of course, and divide their feeding and breeding between the poles and the tropics.

Equally striking to the marine scientist, however, is the variousness of the oceans. Each sea embraces several or indeed many distinct environments, each of which occupies a discrete *zone*. Some of these zones also vary markedly with time – through the day; with the tides, and therefore with the phases of the moon; by season; and sometimes in cycles of several years (see El Niño, overleaf). And superimposed on all these variations in space and time are the more erratic influences of currents and of the influx of rivers.

In short, patchiness, in space and time, is as much a feature of the oceans as it is of land; indeed, 'patchiness' is a great principle in ecology – though it is rarely singled out as such. Each oceanic zone has its characteristic creatures – sometimes a huge variety of different types, and sometimes only a few; but many creatures, of all kinds, spend part of their lives in one kind of environment, and part in another. And because there are so many different ways of making a living in the oceans – so many permutations of habitats – there is a correspondingly huge variety of creatures; and many creatures take quite different forms, and live in quite different ways, at different stages of their lives. There are far fewer species in the oceans than on land, however, because there are no marine equivalents of the forest trees; and it is the trees that provide such a myriad of habitats for land-based creatures.

Let us look, first of all, at the zones, and the reasons for them.

The ocean zones

There are three basic reasons why the oceans are so clearly zoned: topography, light, and heat.

Topography
The outer skin of the Earth – the crust – consists of several, slowly shifting plates of dense rock. Floating on these plates, and covering about 30 per cent of them, are masses of lighter rock, which form the

El Niño

For most of the time the west coast of South America is bathed in relatively cool waters – around 24°C. Minerals well up from the depths as the ocean strikes the land, which phytoplankton use as nutrients, and vast shoals of anchovies (anchoveta) live on the plankton, to provide one of the richest fisheries in the world. But roughly every five years a tide of warm water (up to 30°C) flows down the coast from the equator; the nutrients in this water have already been removed by equatorial plankton. The anchovy fishery then fails. The local fishermen suffer, and so do the seabirds. This warm current generally flows around about Christmas time (which in the southern hemisphere of course is in summer) and so the local people call it 'El Niño', which means 'the little child', referring to the Christ child.

Research during the International Geophysical Year of 1957–58 showed that El Niño was not a local phenomenon. The warm tides in fact flowed right across the Pacific Ocean, originating (it seems) off the east coast of South East Asia and to the north of Australia. It became clear too, that El Niño coincides with a fall in the trade wind, normally blowing across the Pacific from east to west (from America to Asia), a fall known as the 'Southern Oscillation'. Now we know that El Niño and the Southern Oscillation have enormously widespread effects, typically increasing rainfall in South America, but causing drought in much of Africa, South East Asia, and Australasia. In the El Niño event of 1982–83, 200 000 square km of Argentina was flooded and Peru lost a million dollars' worth of potatoes, while Indonesia, Australia, and southern Africa experienced devastating drought.

Now, research by British scientists of the National Environment Research Council is at last unravelling the causes. During the recent round-the-world expedition of the NERC's research vessel *Charles Darwin*, they found – crucially – that the water in the top layers of the ocean to the north of New Guinea is slightly less salty than the South Pacific, to the north of New Zealand. The difference does not seem great – only about one part in 10 000 – but, when combined with other observations, it could be enough to explain El Niño.

To grasp the whole picture, envisage first the trade wind, blowing from east to west. This literally causes water to pile up in the western Pacific: a veritable hill of water. This hill of water stays relatively still for most of the time, and is steadily heated by the Sun. This in turn, of course, gives rise to massive evaporation, and to high rainfall in South East Asia. Much of this rain runs back into the sea, combining with the rainwater from the great rivers of China. As the water piles up in the western Pacific, it then becomes both warmer and more dilute.

But to the north of New Zealand, the ocean current runs in a giant circle, a gyre. As the sun heats it, the water in the gyre evaporates – but then blows away. As time passes, this south Pacific water becomes steadily more concentrated, which means more salty – as the NERC scientists observed.

This difference in salt content leads to a difference in density, which becomes so marked that the dense southern waters flow into the less dense waters to the north of New Guinea. Then the 'hill' of warm water is released, and begins to flow east towards South America. At the same time, the shift in the distribution of heat energy from the sea causes the trade wind to drop, so there is no force to hold the hill in position. When the flow hits South America, that is El Niño.

Once equilibrium is restored – as soon as the pile of water is distributed, and El Niño has run its course – the discrepancy in salinity between the two areas of the ocean starts to build up again; until,

continents. Between the continents are the oceans.

Only the tips of the continents show above the ocean. Beyond the edge of the present shore-line, and extending out to sea, are the true edges of the continents. The submerged fringe of each continent is the *continental shelf*, which may be a thousand or more kilometres in width (as it is to the south of Madagascar) or have a width of practically zero (as around parts of the Philippines). At the outer edge of each shelf, the sea floor plunges precipitously down into the depths. Typically the present outer edges are at a depth of around 200 metres. The waters above the shelf are warmer, lighter, and more turbulent than the depths that lie beyond, and hence are quite after four or five years, another El Niño is triggered. The energy involved in creating an El Niño is vast, and human beings will never be able to control such an event. With greater understanding of its causes, however, scientists hope at least to provide accurate long-term prediction, so that everyone who is directly affected – virtually everyone in the southern hemisphere – can be prepared.

▼ *The path of El Niño: from the seas off South East Asia, across the Pacific, and down the coast of South America.*

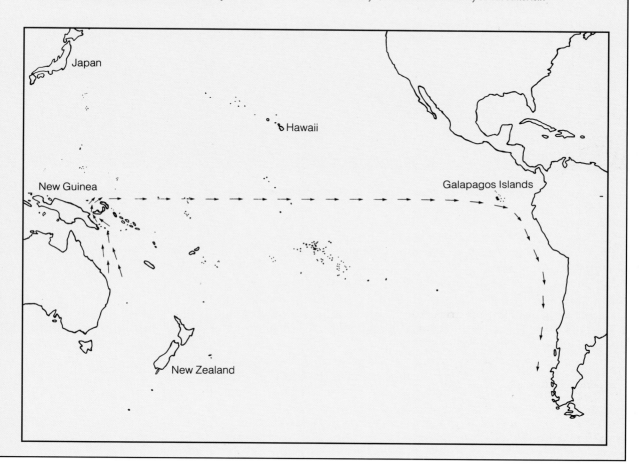

different in character and in the life they support.

Most seas are tidal. At their very edges the sea encroaches far in at roughly twelve and a half hour intervals, and then retreats. So beaches or cliff faces are sometimes exposed, and sometimes covered. These changeable areas are called intertidal zones.

Light
Light entering the sea penetrates at most to a depth of 1000 metres – and since the oceans have an average depth of 4000 m, this means that they are mostly dark, except when lit in the depths by *bioluminescence* (see p.53). There is enough light for photosynthesis

only in the top 100 m or less; the region known as the *euphotic zone*.

Heat

Shallow seas in the tropics may be almost hot – up to 30°C; and the surface layers of the oceans, in the tropics, may be at about 17°C, down to a depth of around 400 m.

But most of the ocean is cold. Of course it grows colder as you move away from the equator, and it is rarely above 5°C at latitudes greater than 50 degrees, north or south. And even in tropical oceans, the temperature declines rapidly below the top 400 m or so; falling sharply to around 4°C at 2000 m.

The general cold, however, is no great drawback. Indeed, to a cold-blooded animal, such as a fish or a crab, low temperatures have several advantages. Oxygen dissolves more freely in cold water than in warm. And if a cold-blooded animal is cold, then it metabolizes less rapidly than if it is warm – and so needs less energy, which means less food. Warm-blooded animals that live in cold water, such as whales, seals, and penguins, insulate themselves with layers of fat (blubber), and reduce the blood flow to the surface. Cold holds few fears for them.

The important point – which is wholly advantageous – is that ocean temperatures are for the most part constant: strikingly so. As we saw in Chapter 2, water has a very high specific heat, which means it takes a lot of heat to change the temperature of the sea significantly; and in cold conditions, the oceans are slow to cool off.

The second reason for the constancy is simply that the ocean is heated from above, by the Sun. Provided water is above 4°C to begin with, it expands as it warms, and so becomes less dense. Thus, as the surface waters of the mid-ocean are warmed, they simply stay on the surface; and (because water is also a poor conductor of heat) they

▲ *The continents sit on shifting plates that form the outer crust of the Earth; and the oceans fill the spaces in between. The water around the continents – on the 'continental shelf' – is shallow. The shelves come to an abrupt stop as the protruding land masses plunge into the depths. Most species of marine organism live on the continental shelf.*

insulate the waters beneath from heating. Water also has a remarkable tendency to expand, if it is cooled below 4°C. So if the deep waters were to become very cold, they would tend to rise to the surface – where they would be heated by the Sun, and join the insulating layers above. In practice, in the absence of storms, there is very little flow of water between the surface and the depths in mid ocean; the warm surface waters stay on the surface; and the deep water also stays where it is, roughly at 3.5°C, all the way to the bottom.

The ocean zones are not, however, quite as rigidly separated as this description suggests – fortunately for living creatures.

Mixing of the oceans

Because they obtain their energy from above, and because water behaves the way it does, there is a tendency for the ocean zones to remain distinct: dark, cold, often almost static waters below; warm waters floating above.

But ocean currents, wind and tides stir the waters. The physical presence of continents complicates the routes of the currents and the tides. For all these reasons, there is some mixing, though it occurs more at some times of year (during the winter storms) than at others. This mixing is very important for ocean life, as we will now see.

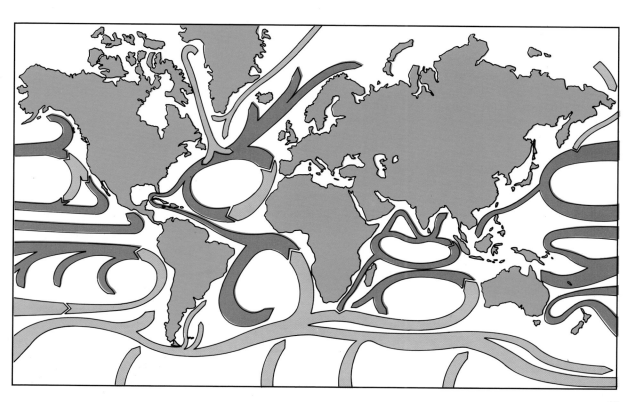

▼ *World ocean currents. Blue: cold currents, red: warm. Several forces combine to mix the waters of the ocean. Currents are one of them. They too have several causes. The overall effect is of enormous complexity.*

Life in the oceans

The most conspicuous feature of life on land is the plants: trees, wherever it is moist and warm enough; grasslands, where it is too dry or too cold for trees; and various intermediate or special states, as in scrubland and bog. In the oceans, large plants grow in only two places. Seaweeds, which are multicellular algae, and submerged flowering plants such as the eelgrass *Zostera*, grow near the edges of the sea. And the calm Sargasso Sea to the north of the Caribbean harbours huge floating mats of weed known as *Sargassum*. But the vast expanse of the oceans seems devoid of plants – at least to the naked eye; and if weed grows conspicuously on coral reefs, then that is a often a sign that all is not well.

The most conspicuous creatures of the oceans are not the plants, then, but animals, ranging from the polyps that make the coral reefs to the great whales and the ocean fish. Yet we saw in Chapter 1 that plants are the source of almost all life on Earth; they alone can create the organic molecules which the animals consume. So how do all those animals live in the apparent absence of plants?

They do not, of course. But plants, as we have seen, can grow only near the surface of the sea, in the photic zone. Unless the sea is very shallow the plants must float. But the surface waters are constantly disturbed by winds. Except in the extreme and anomalous calm of the Sargasso Sea, big floating seaweeds would simply be broken to pieces. The dominant plants of the open ocean, then, are single-celled plants. These are of two main kinds: algae known as diatoms, and flagellates, which have many of the qualities of single-celled animals, but also have chloroplasts, and are able to photosynthesize. Creatures that float near the surface of the seas are called *plankton*. The plants are the *phytoplankton*, and the floating animals are the *zooplankton*. They move by drifting.

Because the creatures of the plankton individually are small, they are not always visible to the naked eye. They are revealed in their millions, though, under the microscope. And the total quantity of phytoplankton is enormous; so much, that it produces 80 per cent of all the oxygen of the atmosphere. The loss of land plants would make a difference to the chemistry of the atmosphere (and of course would be a tragedy). But the loss of these marine plants, which the naked eye may hardly perceive, would spell the end for almost all animals.

On land, the weight of plants far exceeds the weight of the animals that feed on them. The beetles, wasps, and squirrels that feast upon the oak tree are miniscule, compared to the tree itself. In the oceans, this does not seem to be so. There are tiny algae within corals, for example (see pp 56–57); yet life on the coral reef is predominantly animal. So what is going on?

The answer lies in the concept of *turnover*. Forest indeed contains a mass of plant material. But most of that material at any one time is dead. Most of it makes up the timber framework of the trees. Beetles, wasps, squirrels etc feed upon the leaves and acorns of the oak; but a tree that weighs 100 tonnes may produce only a few tonnes of such

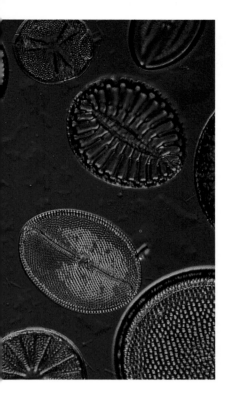

▲ ◄ ▼ *Most marine plants are very small – like these single-celled diatoms of the plankton (above). The large ones, 'seaweeds' such as kelp (above right), generally live close to the shore, where they can anchor themselves. Large floating plants like the Sargassum (left) can live only in seas that are very calm, notably in the Sargasso, north of the Caribbean.*

provender each year.

But the single-celled plants of the phytoplankton need no timber supports. The water holds them up. They are virtually 100 per cent edible. Furthermore, they are extremely efficient. Because they are small, they have a very large surface area relative to their volume, and are able to take in nutrients and light energy from the surrounding water, and excrete wastes, at enormous speed. Thus, in good conditions, diatoms may double their weight in hours. The economy of the forests is reversed. It does not require 100 tonnes of plant material to produce one tonne of edible material per year. Rather, if left alone, one tonne of diatoms could quickly generate millions of tonnes of edible material. At any one time, then, the total quantity of plants in the sea may seem relatively small. But the output – the turnover – is prodigious.

Most of the animals in the sea either feed directly upon the diatoms and flagellates of the phytoplankton, or feed upon creatures that have fed upon that phytoplankton. With this in mind, we can now understand how the various 'communities' of the sea are arranged.

The plankton

Besides the diatoms and flagellates, almost all major groups of animals are represented, in some form, among the plankton. Among invertebrates, the most conspicuous absentees are the insects, which evolved as land animals, and have never returned convincingly to the sea.

Permanent inhabitants include many kinds of protozoa (single-celled animals) and a great many multi-celled animals, many of which undergo a daily vertical migration, from deeper to surface waters, effectively to stay within the same level of illumination as the light varies. Among the most important multi-cellular plankton dwellers are two kinds of crustacean (distant relatives of crabs and lobsters) including copepods and the euphausiid shrimps known as 'krill'. Krill are the principal food of the baleen whales, such as the blue whale and minke. They appear in vast numbers in the Antarctic Ocean particularly in summer (they depend for food upon the blooms of phytoplankton) and it is then that whales like the blue migrate towards the pole to feed.

Floating within or above the plankton, but often as permanent inhabitants, are the jellyfish. Most simply 'swim', pulsing their umbrella bodies to stay afloat. But some, such as the Portuguese man-of-war (*Physalia*) and the velella, have sacs full of gas, which enable them to float on the surface, trailing their poisonous tentacles to catch fish that come to browse in the surface water.

In addition to the permanent plankton inhabitants, huge numbers of animals use the plankton as a nursery – a source of easy food for their offspring – and as a medium of dispersal for the young forms. Animals that do this include fish such as the cod, which of course are highly mobile even as adults; crustacea such as crabs and lobsters; echinoderms such as sea urchins; gastropod molluscs such as winkles, which are mobile as adults but are not able to travel huge distances; and also – crucially – most of the animals which are sessile as adults – that is, remain immobile in one place. Thus the larval forms of corals, sea anemones, most molluscs such as oysters and mussels, worms, and barnacles (which are crustacea) spend time as zooplankton. Animals such as mussels would not be spread so liberally around the world – indeed they could not have evolved at all – unless they had a phase as mobile plankton.

Animals that spend time as larval plankton illustrate two interesting biological principles. First, this phase – though valuable – is obviously highly vulnerable. Cod, for instance, lay about nine million eggs into the plankton. But by far the majority perish, before they are even hatched – or at least before they reach maturity and breed themselves. Indeed, on average, each pair of cod produces only two offspring of opposite sexes that survive to breed in the course of their lives, for if it were not so, the population of cod would grow and grow, which it clearly does not.

The second striking principle is that the larvae that live in the plankton are not, usually, simply miniature or immature versions of the adults. Usually they are quite different in form, just as land-based caterpillars are quite different from adult butterflies. The planktonic larvae are, in short, adapted to planktonic life; those of echinoderms and of gastropod molluscs for example have bands of cilia (whip-like projections of the cell) to drive them along.

All creatures that live in the plankton have to devise means of staying afloat. The many devices range from the gas sacs of *Physalia*, to the liberal laying down of fat. This fat – invariably in the form of oil – is lighter than water, and so improves buoyancy; and also ensures that planktonic creatures, as a group, tend to be highly nutritious.

The waters beneath the plankton contain no plants, but they are full of creatures feeding upon the plankton above, or on the bottom dwellers below, or which are migrating from one rich feeding ground to another. This is the pelagic zone (from the Greek, *pelagikos*, meaning 'of the sea').

The pelagic zone

The pelagic zone contains, in brief, the swimmers. The permanent swimmers belong to four main groups: the squids (which are cephalopods – molluscs – related to the octopuses); the bony fish, most of which belong to the group known as the teleosts; the sharks – which are quite different from the teleosts, though they are also commonly referred to as fish; and the cetaceans, which are the whales, dolphins and porpoises and are, of course, mammals. There are a few other permanent pelagic denizens which ecologically are less important, including the beautiful nautiloids – ancient cephalopods, which still retain the primitive, gas-filled shell; the turtles and sea-snakes; and the herbivorous sirenians – the dugongs and marine manatees of the tropics, which (probably!) are distantly related to the elephants.

The pelagic zone also includes a few creatures that are land-based, but none the less make a significant contribution to the marine ecology: birds such as the puffins, gannets, boobies, auks, terns and penguins; the pinniped carnivores – seals, sea-lions, fur seals, and walruses; other carnivores – the sea-otter and the polar bear; reptiles such as some marine crocodiles and the marine, herbivorous iguana of the Galapagos Islands.

Of the pelagic animals, some fish browse directly upon the plankton, as do the baleen whales, which concentrate upon the euphausiids and copepods. After that, the food chains (see p 110), can be very long; fish such as sharks feeding upon fish, which feed on fish, which feed on fish that feed on plankton. Some pelagic creatures specialize in feeding upon animals near the bottom. Thus the sperm whale makes prodigious dives from the surface, where it breathes, to depths of three kilometres or more, where it feeds mainly upon bottom-living squid. Grey whales dive almost as prodigiously, to filter mud at the bottom and strain out the tiny clams within it.

The pelagic animals generally try to achieve 'neutral buoyancy', to enable them to stay at the required depth without effort; and many have devices to alter buoyancy. Sharks have huge oily livers, which increase buoyancy, but in general they are obliged to keep swimming to provide enough lift to avoid sinking; though some species, in practice, are happy to rest on the bottom. Most teleost fish are equipped with gas-filled 'swim-bladders', with which they can adjust their buoyancy to the particular depth. Some have lost the swim-bladder, however. The mackerel is one such.

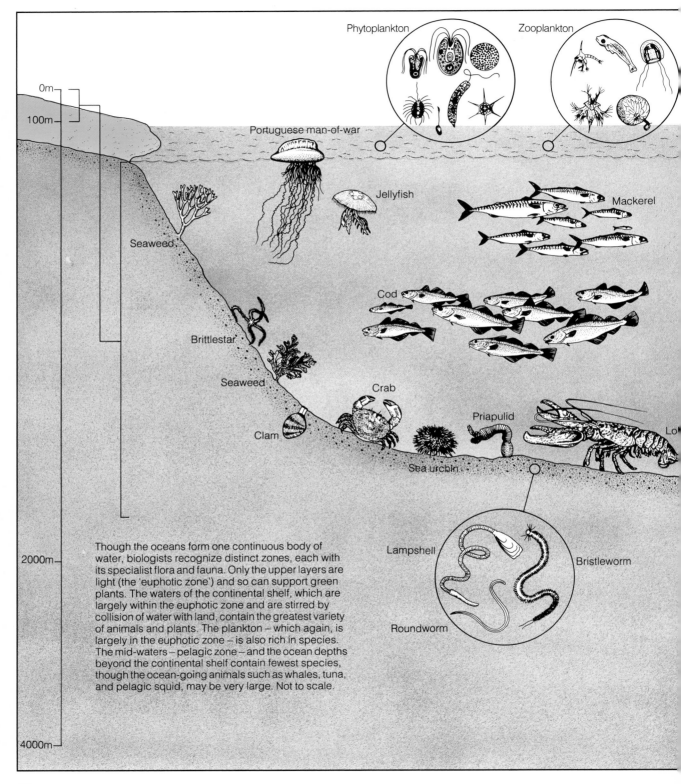

Phytoplankton

Zooplankton

Portuguese man-of-war

Jellyfish

Mackerel

Seaweed

Cod

Brittlestar

Seaweed

Crab

Priapulid

Lo

Clam

Sea urchin

0m

100m

2000m

Though the oceans form one continuous body of water, biologists recognize distinct zones, each with its specialist flora and fauna. Only the upper layers are light (the 'euphotic zone') and so can support green plants. The waters of the continental shelf, which are largely within the euphotic zone and are stirred by collision of water with land, contain the greatest variety of animals and plants. The plankton – which again, is largely in the euphotic zone – is also rich in species. The mid-waters – pelagic zone – and the ocean depths beyond the continental shelf contain fewest species, though the ocean-going animals such as whales, tuna, and pelagic squid, may be very large. Not to scale.

Lampshell

Bristleworm

Roundworm

4000m

Krill

Dolphin

Shark

Grey whale

Squid

Octopus

Plaice

Nautilus

Tuna

Squid

Anglerfish

Deep sea shrimp

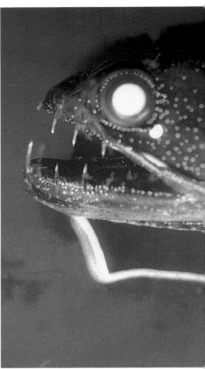

The bottom dwellers

Many animals prefer to spend all or some of their time near the bottom. A huge variety of bivalve molluscs such as clams, and many bottom-dwelling worms, bury themselves in the mud, while others are attached to rock. Some of these then live as filter feeders, passing water through or over their bodies, and extracting organic materials from it with trails of mucus propelled by cilia (which is what mussels do); and others (like many worms) are deposit feeders, taking mud into their guts and extracting nutrients from it.

Many other creatures feed on the animals that feed in this way, or upon the animals that feed on them. These include the mobile and predatory crabs and lobsters, the echinoderms – starfish, brittle-stars, sea-urchins and sea-cucumbers; molluscs such as the predatory whelk, and octopuses. Both the main groups of fish, too, have evolved bottom living forms, which are often flat: like the bony plaice and flounders, and the rays, related to sharks.

The depths

The bottom dwellers mentioned so far tend to prefer the shallower waters of the continental shelf. In deeper waters, beyond the shelf, dwells an astonishing range of creatures adapted to great pressures, extreme darkness, and a fairly sparse and erratic food supply.

Thus, many deep-sea fish have puny bodies (presumably an adaptation to the meagre food supply) but they have huge heads, jaws, and teeth, as their prey – when it arrives – tends also to be large.

▼ ▲ *Life in the ocean depths poses many special problems, requiring special adaptations. Deepwater creatures are few and far between and so have difficulty finding mates. In anglerfish (above left) the problem is solved because the males are parasitic and attach themselves permanently to the females. Many deepwater creatures stay in touch with each other or lure their prey through bioluminescence: pigments that generate light. This* Astronesthes *(above) has been photographed under ultraviolet light.*

Many creatures of all kinds – worms, molluscs, fish – are able to generate light by various chemical means; the depths are often illuminated, at least in flashes, by this bioluminescence. Some fish employ bioluminescence to attract their prey, while others, like fireflies on land, use it to recognize potential mates.

Because food resources are scarce in the depths, animals are correspondingly thin on the ground. This raises problems for creatures seeking mates. Some anglerfish have dwarf males, which remain permanently attached, as miniature parasites, to the much larger females.

This, then, is the general arrangement of marine life. But although the sea is overall a stable environment, it is not unvarying. In most of the oceans, as on land, the ease of living and the distribution of life changes through time.

Cycles and seasons in the sea

The creatures of the open sea, then, live either in the plankton; or live in the waters beneath and browse upon the plankton – or live upon the browsers; or they live on the bottom, and wait for the planktonic manna to rain upon them from above; or they feed on the creatures that feed upon the bottom.

It seems to be a one way traffic. The phytoplankton provides, and then is consumed.

This would be all very well, if the plants of the phytoplankton could make do entirely on solar energy, which at least in the tropics is abundant all year round, and on carbon dioxide, which is always available in solution. But of course, like all plants, they also need a constant supply of other nutrients, such as nitrogen (N) and phosphorus (P). And every time a diatom is taken out of the plankton, it takes some N and P with it.

Thus, the food supply from the plankton tends to be seasonal, and for long periods, over vast reaches of the open ocean, it is virtually non-existent. Typically, in temperate waters, the cycle runs as follows. In winter, the storms stir up the water, and nutrients – such as N and P – are brought up from the bottom. In spring, when the Sun begins to shine, the phytoplankton is able to make use of the newly available nutrients, and flourishes. In spring, indeed, there is a planktonic 'bloom' of algae; not a pathological bloom, a sign of sickness, as is sometimes observed in lakes, but a natural re-awakening of life. But the animals then begin to remove the diatoms, and with them the nutrients they contain. By mid-summer the phytoplankton begins to fail. The cupboard is bare. In autumn – as the storms begin again – there may be a re-stirring and distribution of nutrients, and another bloom. But then the dark of winter puts paid to photosynthesis, until the spring comes again.

In some tropical waters there is no mixing of top and bottom waters and vast reaches of ocean remain almost permanently deprived of plankton. The only animals found here are pelagic creatures passing through.

Near the poles, there is more constant stirring, but the long periods of total darkness in winter impose another form of seasonality. In summer, however, when the sun shines constantly, Arctic and Antarctic waters provide some of the richest feeding grounds of the oceans; fewer species than in the tropics, but often with a far greater biomass.

Again, then, we see the patchiness of the marine environment, both in space and time. As we will see below, marine animals have all adapted to this patchiness – and in some cases, as with the great whales, exploit it beautifully.

Some patches where the sea is constantly stirred tend to be particularly favourable to life. They may colloquially be called 'hot-spots'.

Hotspots: the intertidal zone

On the face of it, the zone between the tides seems a poor environment for living things. Sometimes the animals and plants are under water. But at times they are exposed – and subject, apparently, to over-heating (especially as tidal pools warm up in the sun) and desiccation. Regularly they are pounded by the waves, sometimes violently so. If there is heavy rain during their exposed period, the pools are rendered brackish, rather than saline. Yet, as every investigative child appreciates, the intertidal zones often carry a rich variety of animals and plants, from many different groups, and often in large numbers. How come?

The intertidal zones circumvent the two outstanding *dis*advantages of marine living. Because they are in shallow water – or even exposed – there are no problems with light. They receive at least as much solar energy as the plankton.

Unlike the plankton of the open seas the intertidal zones suffer no shortage of nutrients. The sea at the edges is constantly stirred, both by tidal action and by the waves (reflecting air movements). The beaches may also receive nutrients running off the land, especially near estuaries or sewage outfalls.

Also in the intertidal zones, animals find that much of their work is done for them. Mussels, for instance, do not need to go to look for food, as most animals do; the organic debris they feed upon is wafted in by the tides (or the river estuaries). The work done by the tides may be looked upon as an energy subsidy (see p 116).

Furthermore, there is opportunity for multicellular plants to grow, firmly anchored to the rocks, but still exposed to the light; and many animals – sessile mussels and barnacles; anemones, which are almost sessile; and animals that are only modest movers, such as starfish, winkles, and limpets – find a handy substrate and convenient grazing.

So – if only they can overcome the obvious difficulties, inter-tidal creatures find rich pickings. Many of the animals counteract the relentless pounding by being physically tough, and by their shape; limpets and barnacles, shaped like flattened cones, are ideally suited.

▲ *People in temperate countries pay little attention to mangroves: trees that grow at the ocean's edge. But they are extremely important components of the tropical scene; rich in species, and serving as nesting sites and nurseries for birds and fish.*

Others – anemones, winkles, starfish – are affixed in or move into sheltered places, under rocks, and remain cool and reasonably moist and protected from pounding. Seaweeds such as the brown kelp produce mucus-like materials to hold moisture, and in any case can survive desiccation; and many animals, including shore crabs, shelter among their damp fronds.

So the shore-line creatures adopt several different strategies for survival; and animals from very different phyla – notably coelenterates, molluscs, crustaceans, and echinoderms – have found (various) ways of coping with the intertidal zones. And although the number of *species* in the intertidal zone is not necessarily enormous, individuals of those specialists that have evolved to cope are often present in very large numbers. Mussels and various species of barnacle are obvious examples.

Specialized intertidal communities of enormous significance are the mangrove forests of the tropics. Mangroves are trees that grow right at the edge of the sea, held and nourished through characteristic prop roots that grip the soft mud. They are not a specific kind of tree, but a way of being a tree; several unrelated species (black mangroves, white mangroves, etc) are involved. In Florida, the Caribbean, Queensland, and South East Asia they may be extremely extensive, and are important not only for the huge variety of life within them – crabs, spiders, molluscs – but also for the many creatures that visit them for food (herons, egrets) and which breed in the shelter they provide. In particular, mangroves are the nurseries for huge numbers of tropical fish, which as adults live in the open sea. If mangroves are destroyed (as they often are, to create beaches and hotels – as in Miami) then marine ecosystems a thousand miles away may suffer.

Hotspots: the coral reef

Coral reefs are among the natural wonders of the world; richer in species than any other environment except the tropical rain forests.

As with the intertidal communities, coral reefs flourish in regions that just happen to be especially favoured. They build up in shallow waters, off the edge of tropical or sub-tropical continents, where the waters are stirred, and light penetrates. As with mangrove forests, they do not consist of a single species that happens to have exploited a particular niche, but of a whole group of species, not necessarily related to each other, each of which is adapted to the rich pickings.

Indeed, the main thing that the reef-building creatures have in common is that they are sessile, and produce limestone skeletons, and when they die, the skeletons stay behind. The chief reef building animals nowadays are the corals – communal creatures related to the anemones – of which there are many different kinds (not all very closely related to each other). But many others play their part, including tube-living worms, molluscs (including giant clams, whose shells are of massive thickness) and some limestone-producing algae. We know, too, that in the Devonian period, up to 410 million years ago, there were vast reefs off the west coast of Australia that were made by animals of a class that is now extinct, and were quite different from today's corals. The reef is indeed a way of life, not a particular kind of creature.

Some places on Earth, such as deserts, harbour very few species of living things, and the total weight of creatures, the biomass, is also low. Other places may contain only a few species, yet the biomass at least at some seasons can be very high; and this is true of the Arctic and Antarctic oceans. Other places contain a huge variety of species *and* have a high biomass. The two most notable examples are tropical forests (which way outstrip every other habitat) – and coral reefs.

No-one has ever been able to explain precisely *why* the tropical forests and coral reefs contain so many species, but the two habitats do have several obvious features in common. They are both in the 'wet' tropics: no shortage of moisture, light, or heat, and no interruptions by winter. They also provide three-dimensional environments of enormous physical complexity, and hence contain numerous different niches, each of which might be exploited in many different ways, and in many different permutations.

Like the oceans in general, healthy reefs seem remarkably free of plant life. Again, though, this is largely deceptive. In particular, both corals and giant clams harbour unicellular algae which live *within* their body cells. The corals (or clams) provide the algae with a safe harbour in the sun, and with nutrients such as nitrogen and phosphorus; and the algae provide the animals with much of their food. Giant clams rest with their shells slightly apart, not because they are waiting to trap unwary divers, but so as to expose the algae within the fleshy mantle. The close relationship between corals and algae, or giant clams and algae, is a symbiotic one, from which both partners benefit.

Coral reefs – and the creatures within them – may grow remarkably quickly. The Great Barrier Reef, which extends along the entire

▲ ▼ *Australia's Great Barrier Reef (below left) consists of thousands of coral islands, stretched along the entire coast of Queensland; yet it has all grown up in the past 9000 years. Coral reefs contain a greater variety of species than any other habitat except for tropical forests. Below is a reef in the Red Sea.*

eastern coast of Queensland, and up to 40 m deep, is only about 9000 years old. And the biggest of today's giant clams were not born before Captain Cook, as the tour guides are wont to claim. Their shells may reach a metre or more across, and a thickness of 25 cm at the hinge, in 50 years.

Management of the sea

Because the different zones of the sea are so interlinked – many creatures making use of several in the course of their lives – the ocean environment is far more vulnerable than it at first appears. Damage to especially favoured areas, such as mangroves and coral reefs, can have far reaching effects elsewhere.

Sometimes the worst effects are not the most obvious. Major oil-spills are spectacular. But smaller spills, less obvious but constant, are having a devastating effect on many of the reefs of the Caribbean and the Red Sea. Coral reefs, too, may suffer from input of nutrients – for example from sewage or agricultural run-off – just as earthbound ponds may do. If too much nitrogen falls upon them, then multicellular seaweeds start to grow on them. The algae that live within the coral polyps are then shaded out, and the coral dies. As they are beaten by the waves the exposed limestone skeletons are broken up, and the sand thus formed smothers yet more areas of reef. It has been suggested, too, that pollution by nutrient is responsible for the outbreak of crown of thorns starfish, which have devastated large areas of Australia's Great Barrier Reef (see overleaf).

Fishing techniques can have a devastating effect on marine life. In the Indian Ocean fish are caught in gill nets, which can be up to 50 km long, and are made of synthetic fibres. These break away in storms and drift, catching not only fish but trapping and drowning seals , dolphins and other mammals. As the fibres of the nets are synthetic they do not break down so this dreadful destruction continues.

▶ *The green dots that fringe the mantle of this giant clam are eyes; which register the incoming light and adjust the clam's gape to the needs of the algae that live within its body cells. This is an outstanding example of symbiosis. The algae photosynthesize, and supply the clam with nutrient. The clam provides the plants with a safe haven in the Sun.*

What to do about the crown of thorns

Australian conservationists these past few decades have been worried and at times appalled by the destruction of the Great Barrier Reef by crown of thorns starfish. They have appeared at intervals in vast numbers, eating the living polyps, reducing wide areas of diverse ecosystem to bare white rock that rapidly crumbles into sand. So great and rapid is the devastation that some have feared that the entire reef might disappear – even though it contains 3000 separate coral islands, stretched 2000 km along the Queensland coast, and covering an area the size of Great Britain.

Biologists have been pressured to find a 'cause' for the invasions, so they can be nipped in the bud. It has been proposed, for example – entirely plausibly – that the 'blame' rests with Queensland sugar planters: that fertilizer might run from their fields into the sea and enhance the growth of marine algae. The resulting bloom would then allow abnormally high numbers of planktonic starfish larvae to survive; which in turn would give rise to a huge population of predatory adults. However, experiments in which crown of thorns larvae have been suspended in cages in various parts of the tropical ocean have not supported this idea.

Many attempts have been made simply to control the starfish. Teams of scuba divers have injected thousands of them with copper. It was even proposed at one point that the reef itself should simply be drenched in copper. Scuba divers can protect only small areas.

Some biologists suggest that the best solution to the crown of thorns 'problem' may be to do nothing – except, of course, to monitor the situation. Indeed, Dr Tony Underwood of the University of Sydney suggests that to control the crown of thorns starfish could, in the end, do more harm to the reef than good.

The wait-and-see school points out, first of all, that the Great Barrier Reef, despite its enormous size and importance, is – in biological terms – rather new; only about 9000 years old. The crown of thorns starfish, on the other hand, is an ancient inhabitant of tropical oceans. In other words, the reef grew up in the presence of the starfish. Unless the starfish really is being enhanced by some change of circumstance (for example by fertilizer from the mainland) then it seems inconceivable that there have not been similar outbreaks in the past. And yet the reef has managed to grow, and has achieved enormous diversity. Perhaps the only thing that is different about the modern invasions is that western scientists and divers are around to observe them. Perhaps such invasions are simply a natural feature of reef development. We do know, too, that the reef regularly suffers enormous damage from hurricanes; and yet it recovers.

Tony Underwood also suggests that the crown of thorns starfish may even do good. He points out, after all, that one of the great features of the reef – one that makes it so exciting to biologists and tourists alike – is its bewildering diversity. Often, he says, ecosystems are at their most diverse when they are shifting from one state to another, as the dominant types are temporarily swept aside, and many other, perhaps more specialist species, take over the vacant niches. Often ecosystems become *less* diverse if they are simply allowed to 'settle down'. The depredations of the crown of thorns (together with those of hurricanes) may promote diversity, by forcing areas of reef from time to time to begin all over again. If this idea is right (and Dr Underwood is not saying it is; he simply points out that it is plausible) then elimination of the crown of thorns could *reduce* the diversity of the reef. This is another way of saying that many present species would go extinct.

What is certainly true is that nature is not as straightforward as we like to suppose. Crown of thorns looks to us like a plague; a disease of the reef. But for many of the reef species it could (perhaps) be a saviour.

▼ *The crown of thorns starfish.*

Ocean life suffers, too, from over-fishing. Blue whales are no longer hunted, but their numbers were depleted to such an extent in the past that present populations (perhaps of only a few hundred individuals) may be too small to recover. In general, modern fishing is far too efficient for its own good, with ships often removing entire shoals in one sweep, far out to sea. Vicious circles are set up. When fish stocks are severely depleted, their output is also reduced – because there are fewer fish remaining to reproduce. So the fishermen fish even harder, to make up the catches they need to keep their vessels at sea. In the North Sea, it is already clear that fish stocks would increase – and therefore catches would go up – if fishing was greatly reduced. The paradox is, in short, that fishermen would catch *more* fish if they fished less!

In general, the aim of fisheries advisers is to establish the *maximum sustainable yield* for each prey species. But this varies enormously from species to species, depending on the animal's habits. Some species, including many fish, are long-lived, and reproduce every year for many years. Others – like the squid, which are the chief prey around the Falkland Islands – live for only one year, then reproduce, then die. Some species grow very quickly; some breed very copiously (as cod can do); but others – like whales – breed very slowly.

It is theoretically possible in each case to work out the maximum sustainable yield, and to determine which classes of the animal – superfluous young ones, superfluous males, old ones – can safely be caught without damaging the stocks. But in some cases the necessary basic information is lacking; for example, the reproductive rate of some whales is unknown. In many cases, too, it is difficult to enforce whatever fishing recommendations are laid down.

A huge problem, though, from the conversationists' point of view, is that it may not be in the interests of fishermen or whalers or hunters of ivory – or at least not in their financial interests – to achieve maximum sustainable yield. At first sight this seems to defy common sense, but the point is easily made with reference to whales. Whaling ships are extremely expensive. Every few decades (or less) they need replacing. One strategy – the one that biologists favour – would indeed be to work out maximum sustainable yield, and stick to it. It may well be cheaper, however, to use the existing ships to hunt for whales as vigorously as possible and if the whales go extinct – well, scrap the ships and invest in some other business! In short, financial considerations and biological considerations will be in step only if the pursuit of maximum sustainable yield brings the greatest financial reward. But unfortunately it may not. Such arguments as this help to expose the notion that conservation can safely be left purely in the hands of the market economy.

One of the outstanding problems for conservationists these days, is to reconcile the different uses to which human beings aspire to put the oceans. How is it possible to exploit coral reefs for tourism, for example, but also use them as a food source?

We will refer to such issues in Chapter 8. Meantime, we should look at life on land.

4 The land

The more that scientists observe the oceans the more intricate they find them to be; yet compared with the land they are straightforward. Of course the oceans are not quite unchanging; but at least they protect their inhabitants from fluctuations in water supply, which on land range from virtually permanent drought, to more than four metres of rain per year. The oceans cushion, too, against extremes of temperature, which on land range globally from −126°C (recorded in Antarctica) to +58°C (one day in Libya); and may vary between day and night in some deserts from little above 0°C, to more than 40.

In the oceans, too, as we saw, the turnover of life can be prodigious, but except in the mangroves and a few other places there are no big and permanent plants; the only complex, permanent living architecture in the oceans is provided by the coral reefs. Contrast that with the virtual permanence and size of the world's forest trees and grasses.

Taken all in all, then, the land is intrinsically *more* intricate than the oceans; it has more variation, from time to time and from place to place. Within any one place, it may offer a host of different living spaces, especially within the roots, bark, branches, leaves, buds, and flowers of trees, of different species and at different heights. For example, in a tropical forest, conditions vary from hot, moist and damp at forest floor level to hot, bright and possibly arid in the canopy. It is presumably because of this innate variability of habitat that there are so many species of organism on land: probably anything between 10 and 70 or more million, compared with maybe only a few hundred thousand in the oceans.

Oceanographers traditionally divide the oceans into zones. Land biologists talk of *biomes*, a general term that embraces each region with its characteristic climate, day-length, topography, flora and fauna.

Climate is determined by all kinds of influences: distance from the equator, direction of wind (which in general tends to be east to west around the equator, because the Earth spins from west to east), the distribution of the land masses and the proximity of water, and the positions and height of mountains.

The nature of the soil depends upon the climate (which breaks down the rocks), upon the kinds of rocks involved (hard, soft, limy, granitic) and partly upon their age.

The vegetation is determined by the climate and the nature of the

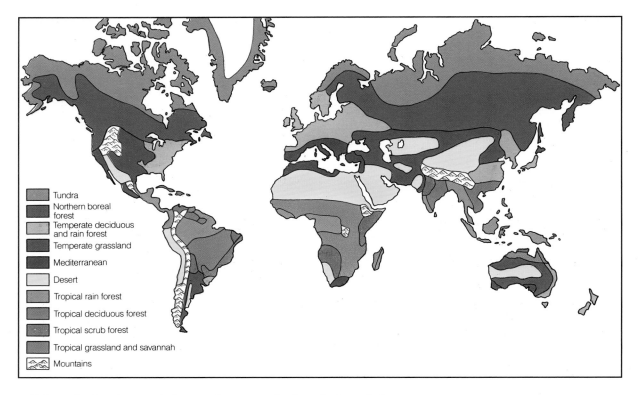

Legend:
- Tundra
- Northern boreal forest
- Temperate deciduous and rain forest
- Temperate grassland
- Mediterranean
- Desert
- Tropical rain forest
- Tropical deciduous forest
- Tropical scrub forest
- Tropical grassland and savannah
- Mountains

▲ *The biomes of the world. Land around the equator is wet and hot. As we move away from the equator the climate becomes drier, then wetter again in mid-latitudes, and then drier again towards the poles. Also, of course, the climate becomes steadily cooler from the equator to the poles. Thus the biomes from the equator to the poles run: wet forest, dry grassland with trees (savannah), hot desert, temperate grassland, temperate forest, and cold desert (tundra). Mountains and islands are special environments.*

soil, but also by time and chance; for example, by what seeds happen to have become established in a particular place in the extreme past. And increasingly, vegetation is influenced by human history – what we decided to plant, or have carried with us by accident. Animal life in turn depends on the nature of the vegetation, and on the climate.

However, the animals affect the plants, and the plants affect the soil (for example, by providing protection against heavy rains, or by breaking up rocks with their roots, or producing acid which dissolves limestone); and the plants also directly affect the climate (for example by re-directing winds, or by transpiring soil moisture into the air).

In short, distance from the equator, winds, lay-out and height of continents, soil chemistry, past history, animals, plants, human history, and time and chance *combine and interact* to produce the particular set of conditions in any one place.

The chief biomes of the world are tropical forest, savannah, hot desert, temperate grassland, temperate forest, boreal forest and tundra. Montane (highlands and mountains) and islands form special biomes. Here we will discuss their biology.

Tropical rain forest

The climate around the equator is both hot and wet. It is ideal for trees; and this, indeed, is the region of tropical rain forest.

In fact, the tropical rain forests are not necessarily quite as tropical, or quite as wet, as their name suggests and as is commonly believed.

▲ *Tropical forests are complex but – in temperature at least, are constant from month to month.*

When the altitude is high (and there is plenty of highland around the equator) they can be positively cold. Sometimes in the forest of Queensland it is very nearly frosty; and George Schaller recalls how the mountain gorillas in the lush forests of Central Africa often sat miserably in the chilling rain. Neither is it always wet; again, the forests of Queensland can at times be very dry (and the tops – the *canopy* – where the leaves are held in the sun all day, must endure conditions that would stress a cactus). Taken all in all, then, the broad expression 'tropical rain forest' covers a multitude of habitats.

What all these tropical forests have in common, however, is their astonishing biological diversity. Charles Darwin, perhaps the greatest field naturalist of all time, was impressed by this when he first encountered the tropics during his journey in HMS *Beagle* in the 1830s. Yet it is only in the past decade that biologists have begun to appreciate the real extent of it.

To begin with, it is not simple in practice – unless you are an experienced naturalist – to see just how varied the trees are. Virtually all tropical forest trees tend to have leaves adapted to deal with the surprisingly harsh and often *dry* conditions high up in the canopy, with extremely varying light, and with very heavy downpours of rain. These physical pressures demand a leaf that is small, dark, and with a point at the end (a 'drip tip') to jettison surplus water like a gargoyle. You can see this typical shape in some house plants, such as weeping fig. But you have to look closely to see that the tropical forest trees that adopt this characteristic leaf form are in fact of

63

literally *hundreds* of different species (see p.124), belonging to scores of different families. In any one hectare of tropical forest you may find trees of one hundred different types and each individual may be widely separated from others of the same species.

More astonishing still, however (in fact quite mind-blowing) is the probable number of small animals, and in particular of insects, that live in the tropical forests. The precise number is unknown; but in the 1970s, two American biologists, TL Erwin and JC Scott, produced a plausible estimate, of around 10 *million*, though it is now thought that the true figure may be around 50 million. Before Erwin and Scott, biologists thought that the total number of species on Earth was probably only two to three million – so they have increased the estimated number by about 25 times. Their work suggests, too, that *most* of the species on Earth – probably around 99 per cent! – live in the rain forests. Their work also confirmed what had long been suspected: which is that the majority of creatures on Earth are insects, and most of them are beetles.

The final, most outstanding feature of tropical forest – not so obvious to the casual visitor, but exceedingly important – is the sheer speed at which detritus (leaf litter and dead animals) is broken down by insects, fungi, and bacteria. There is little leaf litter, so when the trees are removed, the organic content of the remaining soil is low. Some tropical forests grow on rich volcanic soils, and when the trees are removed the land is still good for farming. But other tropical forest soils bake into virtual concrete when the cover is removed. Yet others are like sand. If such soil as this is covered in grass, after a few years of grazing it degrades into desert. What a terrible waste.

Savannah

Country that is too dry for tropical forest but has enough rain to ward off desert contains scattered trees in a setting that is primarily grassland. This is savannah; and it covers much of Africa, India, and Australia. The role of human beings in maintaining savannah is uncertain, but may be profound. In Australia, for example, there is no doubt that the Aboriginal people have done much to encourage grassland these past 40 000 years, by their use of fires, which they light to encourage new growth and attract animals for hunting. One anthropologist has called this practice 'firestick farming'.

Savannah lands (predictably) are dominated by grazers who eat grass, with a sprinkling of browsers, who prefer the softer, more proteinaceous leaves of bushes and shrubs. Africa has a superb array of hoofed herbivores (as well as rodents and primates), including 85 species of antelope, plus various zebras, buffalo, two rhinos, hippos, and elephants. They do not all live on the savannah, of course; hippos prefer rivers and many of the antelope (bongo, duikers) prefer forest, as do elephants when it is available. Between them, though, these hoofed herbivores (ungulates) demonstrate how efficiently vegetation can be divided up and utilized, if nature is given time enough for evolution to work, and a good variety of genes to work

▲ *Africa contains the world's finest savannah: filled (once) with herbivores, grazers and browsers.*

▶ *Termites play an enormous part in the ecology of the world. They remove dead timber, they are wonderful architects, and they generate methane gas, which contributes to the greenhouse effect.*

upon. Thus buffalo, zebra, many an antelope, and the white rhinoceros (which has a square lip) are grazers; while giraffe, okapi, antelope such as bongo and gerenuk, and the black rhino (with its pointed lip) are browsers. Elephants prefer to browse but will graze (as will most browsers) if necessary. Antelope such as impala vary as the year changes; browse when there is browse to be had; grass when not.

India does not have such a broad inventory of antelope as Africa, but it also has deer (which sub-saharan Africa – oddly! – does not). Australia's native animals are mostly marsupial, and – as always – show marvellous similarities to the placental mammals of other continents. Thus the kangaroos are the predominant grazers, and have evolved a stomach for digesting large amounts of coarse vegetation that is similar to the ruminant stomach of the cattle and antelopes. The predominant tree of the Australian savannah ('bush') is the eucalyptus, which is browsed in particular by the koala. The parallel should not be taken too far, however. Giraffe prefer delicate, protein-rich leaves, though they can, when pushed, be much more generalist. Koalas are extreme specialists – they eat very few things besides eucalyptus; and eucalyptus is packed with fibre, tannins, and aromatic ('essential') oils which to most animals would be toxic.

Where there are so many herbivorous animals, we expect commensurate carnivores. Africa has lions, leopards, cheetahs, hyaenas, jackals, and wild dogs. India has all of these except cheetahs (though they did once extend through the Middle East); and its own wild dog, the dhole, replaces the African wild dog. It also has tigers, bears, and wolves, which in sub-Saharan Africa are missing.

The final, most characteristic animal of the savannah is the termite. It gathers detritus (wood, and other forms of cellulose) which it carries into the heart of its giant nests, and uses as a substrate on which to cultivate fungi, on which it feeds. Termites are becoming increasingly important to the whole world. As their gut bacteria ferment cellulose, they generate methane gas, which escapes from the rear ends of the insects. The same effect occurs in cattle, though cattle belch out surplus methane. Methane is one of the principal gases contributing to the greenhouse effect (see p.86). As tropical forest is felled, grassland is spreading to replace it, and termites are becoming more widespread.

Desert and desert scrub

Wherever the conditions for life are difficult, we find, typically, that the number of species is low; and so it is in the world's great deserts that lie just outside the wet equatorial regions, in Africa, Central Asia, North America and Australia. The animals and plants that live in the deserts are all adapted to cope with extreme heat and extreme aridity, and also with discontinuity. They need ways of shutting up shop, or at least of enduring, when conditions are simply impossible. In addition, because there is such a lack of water, the total biomass is low. This means that animals, which live by consuming other organisms, need ways of making do on very little. Reptiles, weight for weight, need far less food than mammals because they do not need to expend food energy in keeping themselves warm. They obtain all the energy they need from the Sun. So it is in deserts (and on islands) that reptiles come into their own.

All the adaptations of the specialist desert plants and animals follow from these requirements. Both animals and plants have ways of trapping water, and of conserving it. Many desert animals, from tortoises to addax and Arabian oryx, have abandoned the need to drink, except perhaps when pregnant. Arabian oryx also lick the dew that may accumulate on rocks and on each others' hair, as the humid air from the Arabian Sea rolls in at night. Tortoises excrete nitrogenous waste in an extremely concentrated form (that is, as uric acid). Many desert plants hide themselves beneath the surface (in the form of rhizomes, bulbs, or seeds) when the going is bad. Cacti, by contrast, demonstrate the arts of endurance. They have abandoned leaves to reduce loss of water by transpiration, and their stems are swollen with stored water. They take in carbon dioxide for photosynthesis by opening their stomata (pores – see p.95) only at night, and then store the carbon dioxide until the following day, when the sun is shining. They may also trap the dew in hairs. Cacti are from the New World – the ones so common in Africa and the Mediterranean have been imported. Many other plants, from several quite unrelated families from the Old World, have adopted the same

▲ *Plants cope with hot dry desert in many different ways. Cacti have deep and spreading roots, store water in their stems, capture water from the air, have a reduced surface area, and practise a form of photosynthesis which requires them to open their stomata only at night (see p.95).*

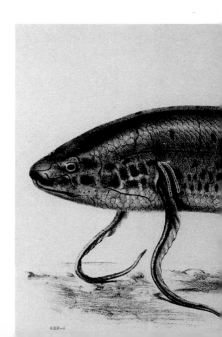

repertoire of tricks, so that a non-biologist would swear that they, too, were cacti. These include some of the Euphorbiaceae, the spurge family, such as *Euphorbia stapfii*; and the 11 members of the remarkable Didieraceae, which are unique to Madagascar. The octopus tree, *Didierea madagascariensis* has spiny, swollen, upright stems like – well, an inverted octopus. When two or more creatures that are unrelated all evolve similar ways of coping with a particular circumstance, this is called *convergent evolution*. The similarities between the cacti, some euphorbs, and the Didieraceae provide a remarkable example of this phenomenon.

Animals cope with discontinuity in various ways. In extremely dry periods, the lungfish of Africa just bury themselves in the mud at the bottom of their temporary ponds, and wait for as long as it takes until it rains again. No one is exactly sure where the reptiles of Australia disappear to in dry periods. Australia's wedge-tailed eagle does not breed till it is six years old, and then breeds only when conditions are good. Australia's marsupial mammals give birth to their young when they are still in fetal form, and – unlike most placental mammals – are able to abort them as soon as conditions turn harsh. In some deserts, as in Australia, the dominant predators are reptiles – snakes and lizards, like the mighty goanna – which can live out periods of hardship with very little sustenance.

Desert animals also adapt their behaviour to the conditions. The mammals in particular stay in burrows during the day, and emerge only at night; and reptiles, too, hole up in the heat of the day.

Temperate grassland

Roughly midway between the equator and the poles, and in the middle of continents, the summers may be too dry for trees to grow, and the winters too cold. Grasses prevail instead. In North America, the temperate grassland is called the prairie; in Asia, it is the steppe, and in South America, the pampas.

Grazers are the most conspicuous native fauna; the larger ones in

▼ *The African lungfish copes with the dry heat of summer by interring itself in the mud.*

Proc. Z.S. Reptilia. XI

general are able to cope with vast amounts of coarse grass, while the smaller ones tend to be slightly more delicate feeders. In North America the mighty bison once roamed in tens of millions. It was hunted by the Indians, but was finally almost wiped out by Europeans in the nineteenth century, partly as a strategy to drive the Indians off the land. The Eurasian steppes had its own bison, the wisent, which was almost extinct in the early decades of this century but has been rescued by captive breeding, and the small saiga antelope, with its peculiar dust-trap of a nose, in shape like an elephant seal's. The pampas has the guanaco and the pampas deer, with the omnivorous, flightless rhea (the South American ostrich) making a trio.

Smaller animals on the grasslands prefer to hide away when they are not feeding; so the prairie dogs, which are rodents, live in vast, colonial underground burrows in North America (with burrowing owls for occasional company), and the sousliks do the same in Asia. Small animals that live in wide open spaces tend to be highly colonial (meerkats of Africa are an example); and they also tend (*vide* meerkats, sousliks, and prairie dogs) to stand upright, scanning the sky for air attack from hawks and eagles.

The rate of breakdown of organic detritus is only modest in the temperate grasslands, so that it accumulates in the soil. It is therefore all too easy to convert to arable farming and the prairie in particular has been reduced to fragments between the expanding farms. Unless

▼ *Prairie dogs from North America are typical grassland rodents, living in communal burrows, constantly on the alert for hawks, snakes, and (till recent years) for black-footed ferrets.*

ploughed grassland is carefully managed, however, it rapidly erodes in the hot dry summers and the powerful winds – as happened in the dust-bowl of the United States in the 1930s. Only with permanent ground cover can the soil be kept intact.

Temperate and boreal forests

From beyond the dry regions of the sub-tropics, stretching north to the tundra and south to Tierra del Fuego, the natural world is shared between forest (wherever there is sufficient water) and grassland.

These temperate forests thus cover a broad span of latitudes, with a wide variety of climates. From the low latitudes to the high, broadleaved trees (which are flowering plants) share the honours with conifers. Where the winters are warm, both the broadleaves and the conifers tend to keep their leaves in winter, as with the evergreen oaks and cypresses of the Mediterranean (though the swamp cypress of Florida is unusual among conifers, in shedding its leaves). Where the winters are cold, the broadleaves contrive to save water in winter, and reduce wind and frost damage, by being deciduous and shedding winter leaves, while conifer leaves tend to have thick cuticles, to be needle-like in form, and are able to withstand the winters. Some conifers, though, such as the larch, have adopted the deciduous habit too.

The natural fauna of temperate forests includes the complete range

▼ ◄ *Temperate forests pose the problem of seasonality. Life is interrupted by winters.*

of feeding types, from herbivorous and detritus-eating insects and mites, through to roe deer, moose, wood bison and various pigs among the herbivores, to bears, wolves, and cats such as the wildcat, lynx, and puma, among the big carnivores.

The contrasts between temperate and tropical forests are striking. The main difference is in numbers of species. Whereas tropical forests may contain hundreds of species of trees, temperate forests typically contain only a few species (oak with ash, yew with beech) or even only one (as in the redwood or cedar forests of North America). And even though each individual temperate tree may support a host of other species (scores of different insects on an oak, for example) the total list of temperate forest animals is also many hundreds of times lower than in tropical forests. Note, however, that most species in tropical forests live over a far smaller area – that is have a far smaller *range*, then the average species in temperate regions.

In addition, life in tropical forests is usually not strikingly seasonal (though some suffer dry periods and wet periods too), and trees fruit and flower at intervals all year round. Temperate trees are highly seasonal, and the animals must be too.

Probably the seasonality and the number of species are related phenomena. We may argue that temperate animals and plants must 'concentrate' on coping with winter, and that only a few manage to do so; or that because life is so seasonal, temperate animals and plants are thrown into greater competition at specific times of year, than in the tropics. But such explanations are of the kind that scientists call 'arm-waving'. The deep reasons for the difference in numbers of species are unknown – assuming, that is, that there are deep reasons.

A third marked contrast, of enormous ecological and economic significance, is in leaf litter. In temperate forests, fungi and detritus-eating insects cannot work as quickly as in the tropics, and leaf litter builds up. When temperate forests are cleared, the remaining ground is rich in organic matter, which means that it can withstand repeated ploughing, and supports good crops. Much of the best lowland farming areas of Europe are former forest. But detritus is given no time to accumulate in tropical forests.

Tundra

Tundra describes the open country to the north of the boreal forests that is too cold for trees. There is little tundra in the southern hemisphere (where there is very little land at appropriate latitudes) and only a narrow strip in Europe, which is warmed by the Gulf Stream; but there are vast tracts in the north of Canada and Alaska, and the northern shore of Asia.

Cold predominates: the mean temperature in some months in the North American tundra may be −30°C, and in Siberia an astonishing −50°C; and in summer, the ground may remain frozen solid ('permafrost') except for the top metre or so. The tundra is dry, too; precipitation is often only 150 to 300 mm per year. Usually there is enough water, however, for evaporation is low as well, and the

▶ ▼ *Adapted in their different ways to the dry and cold of the tundra; musk-ox and opposite-leaved saxifrage.*

plants do not grow rapidly. But some plants, like the opposite-leaved saxifrage, are adapted to extreme drought. There is hardly enough unfrozen soil for trees to form proper roots – especially as they must also withstand powerful winds. The plants are tough perennials, less than a metre high. Britain has a little tundra, in the north of Scotland, not least in the beautiful, wet 'flow country', of Caithness and Sutherland. There, the tallest 'tree' is the dwarf birch, *Betula nana*, a mere 15 cm in height. However, the summer days are long at high latitudes, and even though the radiation is somewhat weak (because the sun is low in the sky), life can flourish.

The tundra animals adopt various kinds of economy. Some are permanent residents. Of these, some, including many rodents, survive the winter by hibernation, while the musk ox, like many hoofed animals in harsh climates, loses much of its appetite when there is no food to be had, and survives largely by endurance, living off its body reserves. The invertebrate residents pass the winter in dormant phase, often as eggs. The small animals (invertebrate and vertebrate) reproduce quickly when the going is good; and as they multiply in spring, the resident carnivores (Arctic fox, snowy owl and rough-legged buzzard) flourish too.

In summer the resident fauna is reinforced by migrants from the south: the reindeer in Asia (called caribou in North America), followed by wolves and brown bears. Wolverines (which are really giant weasels) also come in.

Another tundra strategy is to make use of the sea. Thus the resident polar bears live on seals and fish, while the Arctic tern lives a life of permanent summer and almost perpetual daylight, migrating from pole to pole every year, to feed on small fish.

Islands

Islands are in many ways special: so much so that ecologists have coined the expression 'island biogeography' to describe their particular features.

Because islands are smaller than continents (by definition) the populations of each kind of animal are also bound to be small. But if animal populations are too small, then they simply die out. It follows, then, that the number of different species that can live on any one island is also much smaller than in an equivalent area in equivalent terrain on a continent: there just is not enough resource to maintain sufficiently large populations of lots of different kinds of animal. On many islands, though, the lists of species are enhanced by mobile visitors, which can also live elsewhere; the Galapagos Islands, for example, in the Eastern Pacific, have one kind of penguin that lives nowhere else, but also have three kinds of boobie bird, plus frigate and tropic birds and brown pelicans, which live throughout the warm oceans.

On the other hand, as islands are (by definition) separated from other land, and in particular are cut off from continents, they tend each to evolve their own particular sub-species, at least of creatures that do not easily cross open sea. Thus each of the different islands of the Galapagos and the Seychelles and Mauritius once had their own sub-species of giant tortoise.

Freshwater fish and other species are effectively island creatures, as lakes and upper reaches of rivers are isolated from neighbouring lakes, unless there is flash flooding. So, for example, although there is only one species of brown trout in Northern Europe there are more than 50 distinct types which have evolved as the populations have been isolated. Lough Melvin in Ireland has three sub-populations of the brown trout – Ferox, Gillaroo and Sonaghen – and although these feed together in the same lake they return to separate rivers to breed as adults and so the different sub-populations are maintained.

Because the total numbers of animals on islands tend to be small, large predators in general find it impossible to exist at all – except, of course, for creatures such as seals and sealions, which feed at sea and come ashore only to breed or bask. Native island predators, then, tend to be lizards and snakes, with the occasional bird of prey such as the Galapagos hawk.

Because of the lack of predators, island animals are often extremely tame. This is why boobies have their name (from the Spanish 'bobo' meaning 'clown'). They fed on the flying fish thrown up by pirate ships and whalers – then perched on the rigging, only to be removed at leisure for the cooking pot!

Partly because of the lack of predators, many island birds have lost

▶ *The ecology of islands is in many ways special: few species; no large mammalian predators; large versions of creatures that are generally small, and small versions of creatures that are generally large. Many birds lose the power of flight – for there are no large predators to make it worthwhile.*

the ability to fly; after all, flying requires enormous energy, and on islands, those that do take to the air could in theory be blown out to sea. Thus Mauritius had a flightless pigeon, the dodo, and there were related forms on nearby Rodriguez. Galapagos has the world's only flightless cormorant, and New Zealand has an almost flightless ground parrot. New Zealand also has the flightless kiwi (its national symbol) and was once dominated by the flightless moas, of which there could have been about 25 species. One of these was four metres high, the tallest bird that ever lived, and was finally killed off (tragically) only in the nineteenth century.

Strange things happen, too, to the size of animals. In general,

animal size is a compromise. There are theoretical advantages in being large: large males can compete more easily for mates, and large females can bear larger (and therefore more robust) offspring, or lay more eggs. But there are also advantages in being small, because small animals need less food, and tiny creatures such as mice can hide from predators more easily than larger ones.

On islands, there is little food; but the art of long-term survival is to maintain enough individuals to sustain a viable population. Hence very big animals of any kind are taboo. Thus on the Shetland Islands to the north of Scotland, there are tiny ponies and sheep; and although these are domestic animals, they illustrate the principle. Even more striking were the elephants that once lived on Malta, and were only about a metre in height. These survived until about 8000 years ago, when they were probably wiped out by the first human settlers.

On the other hand, in the relative absence of predators, there is no advantage in being *very* small, like a continental mouse. The theoretical advantages of greater size begin to prevail. Accordingly, islands of the ancient Mediterranean also contained giant dormice. If evolution had continued in that vein, the elephants and the dormice might have crossed in the middle! By the same token, the dodos of Mauritius were giant pigeons (though some biologists believe they were closer to rails). Giant tortoises in the past have existed on continents or quasi-continents (both on South America and Madagascar) but in general they are island animals. They illustrate the principle of *allometry*; that an animal's relative *proportions* may change as it increases in size. Because of this allometric effect, giant tortoises in general are unable to withdraw their heads. If the islands where they lived did contain strong-jawed predators, such as hyaenas, their heads would make a crunchy snack.

However, because island animals have evolved in the absence of predators, because their populations are often fairly small to begin with, and because they have nowhere to run to, they tend to be extremely vulnerable when predators are finally introduced and island creatures are particularly prone to extinction. Hence introduced mongooses (plus cats, dogs, and rats) have laid waste much of the native fauna of the Caribbean (which is also threatened by horticulture and tourism); and cats, dogs, and rats have driven several sub-species of Galapagos tortoise and Indian Ocean tortoises to extinction. The moas have all gone – mostly wiped out by the Maoris, who arrived in about the tenth century AD, and have dispelled the myth that hunter-gathering people necessarily live in harmony with nature. Like agricultural people, sometimes they do and sometimes they do not.

Islands are of particular theoretical interest because nature reserves and lakes are also, effectively, islands; limited areas, cut off from similar areas. The animals within them are therefore extremely vulnerable, mainly because the populations within them tend to be too small for safety. Indeed, ecologists speak of 'species relaxation', to describe the steady loss of species after a new nature reserve has been

▲ ▶ ◀ *Island evolution: a giant tortoise and over-trusting boobie birds (from Galapagos), and a flightless brown kiwi from New Zealand.*

created. Conservationists have an enormous job to keep the populations high enough for continued existence, while preventing them from growing too large and wrecking the little bit of space that is left to them.

Montane

Moving from the foot of a high mountain to its peak is very like travelling from the equator to the pole. It gets colder as you go up (by 0.5°C with each 100 m rise), and the plants and animals change accordingly. The difference lies in the rainfall. Often, one side of a mountain is wet, and the the other much drier. There is not the consistent pattern of wet-dry-wet-dry that you find when you move through latitudes.

Mount Kilimanjaro, 5895 m in height and standing practically on the equator in Africa, demonstrates this change in biomes with altitude beautifully. The bottom has tropical forests. These give way first to deciduous forest, then to conifers, and then the trees give out altogether, giving a stretch of low, alpine vegetation, culminating – even though the mountain is on the equator – in permanent snow.

The plants of high mountains have some of the features of desert plants – often succulent, tough leaved, and resistant to water loss; often hiding in winter, and flourishing in spring, like the wonderful flowers of Alpine meadows. But because mountain peaks require specialist flora and fauna, and are often separate one from another, they also to some extent resemble islands; and indeed, they have been called 'sky islands'. The populations that live on these sky islands are cut off from each other, and evolve independently. Before long, there are clear differences between the different populations, so

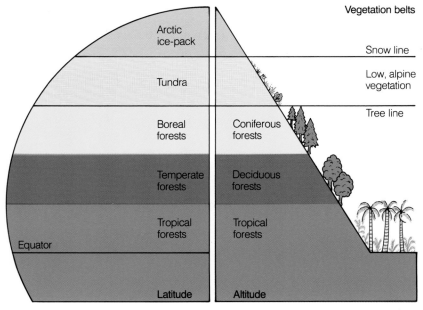

◄ *Mountains in the tropics are like the world in miniature: hot wet forest at the bottom, cold and tundra-like at the top, and everything in between.*

that each one is recognizable as a distinct 'race', or 'sub-species'. This is indeed what tends to happen in groups of real islands. In addition – as on real islands – the total number of species on any one peak is generally low, and many bizarre forms evolve. Mountain peaks, like intertidal zones and estuaries, are to some extent enriched by organic material blowing in from elsewhere – notably in the form of insects, borne aloft, and perishing in the cold.

Artificial ecosystems

Within these main, broad biomes, there are hundreds of specialized systems: saltmarsh and sand-dune; cloud-forest (at the tops of high tropical mountains) and forest-floor streams; ponds and rocks. And in addition to all these again, there are thousands of kinds of habitat – or potential habitat – created consciously or inadvertantly by the activities of human beings.

Human-created habitats have their own special set of characteristics. Commonly, they contain species from several different kinds of natural habitat. None of these species may feel entirely 'at home' in the new place – but together, they may make an impressive species list. Traditional hedgerows illustrate this, with a mixture of species from woodland and open fields. Commonly, too, human activity creates habitats that are in a constant state of flux, which may open the way for pioneer species (such as ragwort and dock – the kind commonly regarded as 'weeds') but also produces rapid successions of species. Thus the traditional coppiced woods, in which chestnut trees were cut close to the base to produce a regular supply of straight sticks, left grassland in between that was sometimes open (when the sticks were harvested) and sometimes shaded (when the crop was grown). Again, the total species list was high.

Humans create niches for wildlife by providing extra pockets of nutrients – comparable to estuaries, where nutrients are brought in naturally from the surrounding seas and landscape. Cities are an example, with their constant supplies of organic detritus enjoyed, for example, by herring gulls near the coast, and by black-headed gulls further inland. In Yellowstone National Park in North America, the grizzly bear population benefited from its raids on the garbage heaps until these were fenced off. Many populations of animals worldwide – foxes, raccoons, coyotes, possums, and even sheep (in South Wales) and polar bears (in Alaska) – are maintained in part by raids on dustbins.

Finally, humans create habitats of a kind that may well occur in nature, but not commonly. Eaves of houses are artificial cliffs for martins; office blocks and ancient churches fill the same function for peregrine falcons in Salt Lake City in the United States, and in Prague. Bats roost naturally both in caves and hollow trees. But they are happy in buildings, too; not simply in the traditional belfry, but also (if left alone) in the lofts of modern houses. The residents of one housing estate in the North of England have taken their resident pipistrelle bats, several hundred of them, under their metaphorical wing; holding street-parties (or at least garage parties) to pool data. It *is* possible to live with wild animals, if we give them a little space.

Often, human beings have created these opportunities for other animals only inadvertantly. Farmers traditionally tried to get rid of creatures in which they had no direct commercial interest, and town dwellers tried to banish moles, squirrels, and foxes from their gardens, and bats from their lofts. Even when people took an interest in wildlife, and visited wild places, they tended to trample on or otherwise disturb the creatures they admired. 'Each man kills the thing he loves', as Oscar Wilde observed.

One of the great challenges now for ecologists, planners, economists and politicians is to reconcile the aspirations of humans with the needs of other species. We need to develop techniques to allow tourists to see animals and plants, without destroying them; to devise systems of farming that deliberately create opportunities for wildlife, and to design our cities with wildlife in mind.

This is discussed further in Chapter 8. For now, we should look at more of the essential components of life on Earth: the energy that comes from the Sun, and which keeps the whole system rolling; and the atmosphere itself, which protects us from too much of that radiation, and provides several of life's essential ingredients.

▶ *Man-made environments can be extremely rich in species – partly because, like this chestnut coppice, they may be constantly in transition.*

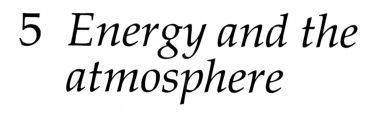

5 Energy and the atmosphere

In this book we have spoken of life as a series of dialogues: between carbon and water; between proteins and nucleic acids; between competition and co-operation. In this chapter we discuss yet another dialogue, between the various forms of energy – radiations – that bombard the Earth from space, and the gases of the atmosphere that envelop the Earth. Human activities, now, are upsetting this dialogue, and we will look at how we are changing the atmosphere, ever so slightly but none the less crucially, in ways that could endanger all of life.

The first participant: energy from space

The radiation that strikes the Earth – or at least gets as far as the outer atmosphere – is of two main kinds.

Some of it consists of streams of charged particles, fragments of atoms that carry an electric charge. These streams are known as *cosmic rays*, and come from various parts of the Universe. Cosmic rays strike against atoms in the atmosphere as they pass through, and thus generate more charged particles and other forms of energy, and it is these secondary radiations that reach the Earth's surface. Individual particles in cosmic radiation have extremely high energy, which implies that they are potentially damaging. But fortunately the total amount that reaches the Earth is not great.

Far more important than the bombardment of such particles is *electromagnetic radiation*. Physicists envisage this in two apparently different ways: either as streams of particles known as *photons* (which are quite different from the particles of which atoms are composed); or as *waves*. It is odd – one of the anomalies of science – that it is possible to look at a single entity *as if* it were either one thing or another, apparently quite different thing. The fact is, however, that some of the behaviour of electromagnetic radiation is best explained as if it were a stream of photons; and some of the properties are best explained as if it consisted of a series of waves.

Photons are often spoken of as 'packets of energy'. Some of these packets have very low energy – and these correspond to radiation with extremely long wavelength; and some packets have very high energy – which correspond to radiation of short wavelength.

Radio waves are electromagnetic waves with a very long wavelength, measurable in metres. Microwaves are much shorter, measured in millimetres. These are detectable throughout the Universe (in low amounts) and were produced originally by the Big Bang: the explosion that gave rise to all the matter and energy in the Universe, 15 billion years ago.

Shorter still is infra-red radiation, and then comes visible light; both measurable in millimicrons or angstroms. Red has the longest wavelength of visible light, and violet the shortest.

Ultraviolet light has even shorter wavelengths, with UVA longer than UVB. X-rays and gamma rays are the shortest electromagnetic waves, with wavelengths less than a 1000 millionths of a centimetre.

Whether or not a particular form of radiation has any effect on a living creature depends upon three factors. The first is the energy that the radiation carries. Thus short-wave radiation, which has more energy, is likely to be more dangerous than long wave. Indeed, both gamma rays and X-rays can be extremely damaging to living cells.

The *dose* received is also important; and this depends largely on the amount of the particular radiation that reaches the Earth's surface. The atmosphere filters out most of the gamma and X-radiation so that the amount reaching ground level from space is virtually zero, and astronomers who seek to measure these short-wave radiations must send their telescopes into space.

The final consideration is whether the particular radiation is absorbed by the living material, and thus passes on its energy to it; or is reflected from the surface, leaving little energy behind; or simply passes straight through, without interacting with the atoms of the living flesh.

The Sun radiates energy over the whole electromagnetic spectrum, but the Earth's atmosphere is transparent only to infra-red, visible, and some ultraviolet light. These then, have by far the greatest effect on living things.

How the Sun's radiations affect living things

Infra-red light reaches the Earth in significant amounts, and its effect on the molecules of both living and inanimate objects is to vibrate them, which means that it raises their temperature. Indeed, infra-red light is chiefly responsible for the general warming effect of solar radiation. Without it, the Earth would be too cold to sustain life. Many 'cold-blooded' animals, from lizards to butterflies, sun-bathe in the mornings to get themselves moving. Lizards visit particular rocks in their terrain that have been heated by the Sun. A few seconds is enough to give themselves a boost!

Ultraviolet light is a mixed blessing as far as living things are concerned. For the most part, it is bad. The photons of which it is composed have extremely high energy, and instead of merely vibrating the molecules of living tissues and heating them, they

▲ *The beneficence of the Sun: an occasional visit to a sun-warmed rock may be all a lizard needs to keep itself warm.*

▶ *Some animals – like bees – can 'see' ultraviolet. Some flowers, accordingly, are ultraviolet-coloured. Left is a flower as seen by the human eye; photographed in 'normal' light. Right, the same flower as seen by a bee, photographed in ultraviolet light.*

infiltrate organic molecules and disrupt their chemistry. UV light seems to have a particular penchant for DNA, and the chemical changes that UV radiation causes in it are manifest as genetic mutations. Doubtless, ultraviolet light has helped evolution to move along by causing some of the genetic variation that natural selection must work upon. In the short term, however, mutation is far more likely to be harmful. One of the immediate consequences is for cells to become cancerous. In humans, malignant melanoma – a cancer of the cells containing skin pigment – is a particular hazard of excess sunbathing. However, many animals utilize ultraviolet light to some extent. Very commonly it provides the energy for the synthesis of vitamin D (tortoises kept as pets in northern latitudes may find it difficult to breed because they are D-deficient). Some animals, too, can 'see' at least the longer wavelength forms of ultraviolet light. Many flowers appear to bees to be ultraviolet-coloured even though to us they appear to be a quite different colour.

No animal or plant can withstand ultraviolet radiation in more than modest amounts. Corals, which must remain close to the surface of the water because they contain symbiotic algae which photosynthesize, also tend to contain specific pigments to protect them from ultraviolet light. Some fungi contain similar compounds. Manufacturers of suntan lotions are now expropriating and synthesizing these materials.

Living things use the energy of visible light as a source of

information about the world. Many plants, for example, are sensitive to the length of day, and have geared their lives accordingly. The time of germination, flowering, fruiting, or leaf-fall and general senescence, may be determined by day length – for day length, after all, is a much more reliable indicator of the time of year than temperature is, for example. Many animals, from sheep to starlings, breed only at certain times of year – again depending on day length.

Many creatures, too, use visible light to tell them what is happening in the world immediately around them. Even earthworms have light-sensitive cells in their skin which tell them whether it is light or dark. In many other animals, such cells are overlain with lenses, to focus the light, and attached to nerves that converge on some central brain; a set-up that lays the foundations of true vision.

In addition, some bacteria and all green plants have evolved ways of using light as a source of energy. In effect, they convert the energy of the electromagnetic radiation into chemical energy. That is, they use light energy to split water, and extract hydrogen; and then combine this hydrogen with (low-energy) carbon dioxide to produce high-energy organic molecules: molecules which in turn act as fuel and as the raw material from which is fashioned almost all living flesh, both animal and plant. The processes by which light energy is thus harnessed are collectively called photosynthesis.

We will look again at photosynthesis on page 93. First we should discuss the atmosphere, and how it intercepts and modifies the various radiations.

The second participant: the atmosphere

In second form science (at least in the author's day) we are taught that the atmosphere contains roughly four-fifths nitrogen; one-fifth oxygen; and traces of carbon dioxide, water vapour, and 'rare' or 'inert' gases, such as neon and xenon. So indeed it does. The rare gases take very little part in life's processes; and we need not discuss water vapour at too great length, even though many desert animals and plants, and some epiphytes (plants that grow on the surface of other plants) derive much or most of their water from water that floats to them on the air. Modern living things have, however, adapted to the presence of atmospheric nitrogen, oxygen, and carbon dioxide; and each of these is pressed into service, in ways that we will discuss later.

The importance of an atmospheric component to living things is not related simply to its concentration. In particular, the air now contains only about 0.05 per cent of carbon dioxide: a trace indeed. Yet all but a fraction of the carbon in living things originates as carbon dioxide in the atmosphere; as we will discuss below, it is turned into organic molecules by plants by the process of photosynthesis.

The concept of *turnover* can be more important than that of mere *amount*. We met this same idea in Chapter 3 (p.46), when discussing the amount of plants in the sea. In places there may appear to be almost none; yet we know that huge numbers of diatoms are produced, but that most are rapidly consumed.

So it is with many gases. In addition to carbon dioxide, the atmosphere also contains trace amounts of methane, ammonia, hydrogen, oxides of sulphur and oxides of nitrogen. But the fact that each of these gases is *present* only in small amounts at any one time, is deceptive. Each of them is being pumped into the atmosphere in vast amounts: and the reason they are not found in vast amounts is that they are extremely reactive or soluble in water, and so are quickly converted into some other compound, or washed out by rain. Thus methane is quickly oxidized to carbon dioxide and water, and ammonia is quickly washed out. *En route,* however, these gases that are just passing through can in various ways be enormously influential, as we will see.

However, the chemistry of the atmosphere is far more complicated than this list of ingredients suggests, for many of the atmospheric gases react with each other to produce yet more compounds, some of which are extremely short-lived (and therefore difficult to measure). Here is where we get the first hint of dialogue between the atmospheric gases and the radiations from the sky: for the radiations provide the energy for many of these reactions to occur. Thus, for example, the various oxides of sulphur and nitrogen may be turned briefly but importantly into radicals (charged components) of acids, sulphuric and nitric, which tend to make rain more acid. Of vital importance is the temporary conversion of oxygen to ozone to create the ozone layer, which in turn protects us from too much radiation.

Radiations cause chemical reactions to occur in the atmosphere. But as they do, they give up some of their own energy; so the atmosphere in turn modifies the radiations. This is what makes the interaction a 'dialogue'. Two interactions that have now become of critical importance are the effect of the ozone layer on the passage of ultraviolet radiation; and the effect of carbon dioxide and various other gases on infra-red radiation. Both of these are discussed later.

Particular components of the atmosphere may act as *catalysts*; speeding-up chemical reactions that would not otherwise occur so rapidly. Of particular significance nowadays is the effect of chloro-fluorocarbons (CFCs) on the ozone layer. These man-made gases are present in the atmosphere only in tiny amounts, but because they accelerate the breakdown of ozone, they increase the hazard of excess ultraviolet radiation. CFCs are not in fact short-lived in the atmosphere, as methane is, for example. Indeed, any one molecule of CFC may remain aloft for decades. This is a characteristic of catalysts: that they cause other reactions to happen while remaining intact themselves. The other characteristic is that their influence seems out of all proportion to the amount present: a little goes a long way.

It has become clear from all that we have said that the composition of the atmosphere is not fixed. It can and does change, even from year to year; and may change appreciably from decade to decade. This concept is crucial to an understanding of all biology for, as we

Pollution of the atmosphere – the greenhouse effect

Various atmospheric gases trap infra-red radiation as it emanates from the Earth's surface, and thus prevent excessive cooling when the sun is not shining. Thus, these gases act like the glass in a greenhouse, which similarly inhibits the loss of infra-red. But it is important that these 'greenhouse gases' remain at an appropriate level. If they are too low, then the Earth's temperature would fluctuate more than it should. If they are high, then a planet can become very warm indeed. Venus is exceedingly hot, partly because it is nearer to the Sun than is the Earth, but largely because its atmosphere is so rich in 'greenhouse' gases.

The most important greenhouse gas is carbon dioxide, which is produced naturally by the breakdown (oxidation) of organic compounds; of sugars, as animals respire, and of vegetation and flesh, through fires and all the processes of decay. CO_2 is also produced in huge amounts by burning fossil fuels – which of course are organic molecules built up by photosynthesis millions of years in the past. Roughly half of all the CO_2 produced dissolves in the oceans, where some is photosynthesized by plants, is taken up by trees or is taken up within limestone (calcium carbonate). The other half escapes into the atmosphere.

Other major 'greenhouse' gases include methane and the CFCs (which also, of course, are the chief destroyers of the ozone layer). Methane is produced naturally from anaerobic decay in marshes, in the breath of ruminant animals such as cattle and antelope, and from the rear ends of termites. Output is now being increased as the domestic cattle population increases, and as tropical forest is felled to make way for grassland – which is ideal territory for termites. CFCs are produced only in minute amounts compared to CO_2, but they are extremely powerful greenhouse gases.

The proportion of greenhouse gases in the atmosphere is steadily rising, because of human industry and agriculture. In 1850 the atmosphere contained only 0.028 per cent CO_2: that is 280 parts per million (ppm). Now the figure is 350 ppm. By the end of the next century, it could reach around 500 ppm. CFCs are of course a new phenomenon; entirely man-made. Methane increases as agriculture spreads and intensifies.

The effects of greenhouse are not easy to discern. Other factors can alter the Earth's climate from millennium to millennium and decade to decade, including slight changes in its orbit around the Sun, and the effects of greenhouse are superimposed on these other factors. We know that greenhouse

effects can be enormous, however. Analysis of air bubbles trapped in the ice of the Antarctic at the time of the last Ice Ages, and now buried deep beneath the surface, show that at that time CO_2 levels were extremely low. The Ice Ages, in short, were caused by a greenhouse effect in reverse.

So unless there is something very wrong with our elementary physics, it seems that the world is bound to warm up by about 2°C by AD 2050. This may not seem a great deal. But a *global* cooling of only five degrees or so was enough to produce the Ice Ages. And the increase in temperature will not be felt evenly. It would be less at the equator (perhaps half a degree) and more at the poles – perhaps six degrees; certainly enough to melt a great deal of polar ice.

Some politicians have made light of these effects, suggesting, for example, that Britain would be a pleasanter place if its climate was more Mediterranean. This is naive. Because the poles will heat more than the equator, there will be a tremendous imbalance in energy over the next few decades, which seems bound to produce enormously turbulent and changeable weather, including gales, floods, and late frosts. Neither can we guarantee that everywhere on Earth will be warmer as the greenhouse effect progresses. For example, shifts in energy distribution could cause ocean currents to change direction – and if the Gulf Stream changed course, Britain could be a lot colder. Neither can we yet predict what will happen to rainfall. Britain could be as wet as New Zealand – or a desert. Agriculture will have to be re-thought, because most crops, at present, grow close to their physiological limits; and present day wildlife reserves may prove unsuitable for the plants and animals they contain, as the climate within them changes. During the Ice Ages of the past, we know that animals and plants shifted – the animals simply walking north or south as conditions demanded. But they cannot walk from reserve to reserve without going through cities. Are we going to organize a mass evacuation?

The release of ice from Antarctica and from northern continents such as Greenland will also cause sea-levels to rise. (The ice of the North Pole is floating – and its melting will make no difference to overall sea levels). The loss of land, and in

continued over

▶ *The greenhouse gases: methane (CH$_4$) from cattle, termites, and swamps; chlorofluorocarbons (CFCs) from aerosols (and fridges); and carbon dioxide (CO$_2$) from everything that breathes and burns.*

continued from previous page
particular of coastal cities, farms, and ecosystems, will of course be serious. But in global terms, it will be far worse than is easily imagined. The British scientist James Lovelock has suggested that life is at its richest on Earth during Ice Ages. To be sure, the high latitudes are out of action, because they are covered with ice. But, says Dr Lovelock, the sea levels fall during Ice Ages – and this exposes vast areas of continental shelf in the tropics. It is good to gain so much tropical space – because the tropics harbour far more species than the polar regions!

In practice, the point is complicated by other factors. For one thing, when ocean levels are high (during warm periods) land is more fragmented. Populations of animals and plants are therefore more fragmented too – and so they are more likely to diversify. In addition, when sea levels are high, more of the continental shelf is covered, and when the land is divided into islands, the total coastline is longer. Both the continental shelf and coastline are in general rich in species. So perhaps the *total* inventory of species during warm times is, in practice, as high or even higher than in cooler times. Dr Lovelock's suggestion is salutary none the less. At least it shows that warmth is not necessarily the biological bonus that some commentators have chosen to suppose.

Steps are being considered worldwide to curb CO_2 and CFC emissions, but the hope of reducing the greenhouse effect seems fairly forlorn. The Third World has yet to industrialize; methane continues to accumulate as agriculture spreads; and the loss of tropical forest reduces CO_2 absorption. Besides, the temperature would continue to rise over the next few decades even if greenhouse gases stayed as they are, simply because the world is so slow to warm up, and has not yet responded to what is there already.

The greenhouse effect could have some pleasant side-effects. Yorkshire for example may regain the vineyards it had during the warm spell of the fifteenth century. Taken all in all, though, it must be seen as the greatest of all the pending threats to present day life.

▼ *The impact of the greenhouse effect will not be simple and uniform. The global warming and the accompanying changes in rainfall are likely to be highly variable, and unpredictable in detail. This is just one of the possible scenarios.*

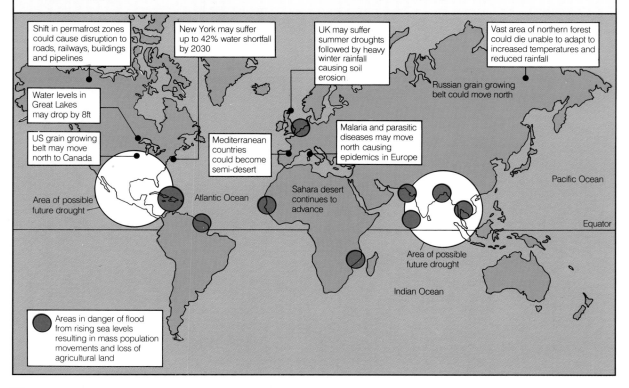

Shift in permafrost zones could cause disruption to roads, railways, buildings and pipelines

New York may suffer up to 42% water shortfall by 2030

UK may suffer summer droughts followed by heavy winter rainfall causing soil erosion

Vast area of northern forest could die unable to adapt to increased temperatures and reduced rainfall

Russian grain growing belt could move north

Water levels in Great Lakes may drop by 8ft

US grain growing belt may move north to Canada

Mediterranean countries could become semi-desert

Malaria and parasitic diseases may move north causing epidemics in Europe

Pacific Ocean

Area of possible future drought

Atlantic Ocean

Sahara desert continues to advance

Equator

Area of possible future drought

Indian Ocean

Areas in danger of flood from rising sea levels resulting in mass population movements and loss of agricultural land

Pollution of the atmosphere – the ozone layer

Oxygen has accumulated in the atmosphere only in the past three billion years. If the ancient bacteria that lived before that time had been intelligent, they would have recognized it as a very serious pollutant indeed. It is, after all, extremely reactive. It reacts with the organic materials of living flesh. To organisms that are not adapted to it, it is highly toxic.

But oxygen does perform one function that brings benefit to *all* of life; not when it is in its usual form, but when it takes the form of ozone.

Under most conditions, the molecules of oxygen gas contain just two atoms: its formula is O_2. Under various influences, however, including electric sparks and exposure to ultraviolet radiation, some of its atoms group in threes, to form molecules with the formula O_3. This is the gas known as ozone.

In the modern atmosphere, a layer of ozone persists in the stratosphere, about 25 km above the Earth. Even in the pristine atmosphere, unpolluted by humans, this layer appears highly unstable since it is constantly being formed and broken down, mainly by the action of ultraviolet radiation. But because ultraviolet radiation is a constant feature of solar radiation, the ozone layer is maintained. And as it is buffeted by ultraviolet radiation, so it absorbs it; and thus prevents the most harmful, high-energy, high-frequency radiation known as UVB,

from reaching the Earth.

However, because ozone is so chemically reactive, it is also readily destroyed even by small amounts of pollutant gases rising from the Earth. In particular, chlorofluorocarbons (CFCs) are released from aerosols, and are used as coolants in refrigerators, from which they escape. CFCs are broken down in the upper atmosphere by UVB, and release chlorine atoms. Chlorine acts as a catalyst, promoting the breakdown of ozone; and each chlorine atom may destroy 100 000 ozone molecules.

Other gases also catalyse the breakdown of the ozone layer, including nitrous oxide, N_2O, which is released from agricultural fields. Methane, which is in other ways destructive (see p.86), fortunately helps to build up the ozone layer.

It seems highly likely that CFCs and N_2O are already damaging the ozone layer. Satellite observations clearly show an ozone 'hole' over Antarctica, where air-currents perhaps cause CFCs to accumulate; and others could appear elsewhere. It is relatively easy to legislate against CFCs – but they may persist in the atmosphere for 100 years, and so will continue to build up; and it is far harder to suppress N_2O production. Some scientists predict a great increase in skin cancers in humans over the next decades, and the effects on other life forms can hardly be calculated.

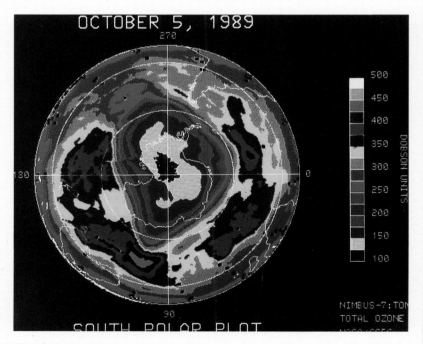

▶ *The hole in the ozone layer is real; shown here over the Antarctic, in a computer-enhanced picture taken by NASA.*

will see, even the broad composition explained in second-form chemistry, with four-fifths nitrogen and one-fifth oxygen, is far from fixed. Oxygen itself has been present in significant amounts in the atmosphere for only about two and a half billion years – just over half the total life of our planet; and there had already been a billion years of evolution before that. We, and all other living things, are adapted to the atmosphere as it now is.

In summary, then, the importance of a particular gas in the atmosphere to life on Earth may be out of all proportion to the concentration of that component. A gas that is present only in a trace – carbon dioxide – is, in practice, the stuff from which all flesh is made. Many 'trace' gases – including some that are unstable, such as ozone – affect the nature and the amount of radiation that reaches the Earth's surface. And some pollutants in the atmosphere act as catalysts, in turn having enormous effects on other components of the atmosphere. If those other components are themselves of a kind that, for example, influences the Earth's supply of radiation, then we see that positively minute additions, at least of some kinds of pollutants, could have profound consequences for living things.

This point is worth emphasizing. Scientists have been warning for at least 30 years that humankind's pollution of the atmosphere was about to have serious effect. But to people who are not trained in science it simply did not seem possible that mankind's puny factories, and even punier cars and refrigerators, could have an effect on the well-being of the entire world. But when you bear in mind the background points (catalysis, the interaction with radiation, the notion of turnover) you see how it is that very small amounts of pollution really could have far-reaching effects.

Pollution of the atmosphere: an upset of balance

Human beings learnt to control fire about 100 000 years ago, causing gouts of smoke and carbon dioxide to be released into the atmosphere. Yet our early fire-raising ancestors hardly qualify as polluters. There have always been fires on Earth, from volcanoes and lightning; and when a plant is burnt it gives out only the carbon dioxide that it itself has taken up in life, so the net effect of burning biomass, at least on atmospheric CO_2, is probably zero.

More serious, probably, was the extensive deforestation carried out by neolithic farmers from around 8000 BC onwards: a deforestation which, for example, denuded Scotland in the centuries before the Romans came. This would reduce the ability of the living world to take up carbon dioxide. Again, though, the effects were minimal. Trees are less important as takers-up of CO_2 than are marine diatoms, and most of the world's trees grow (or grew) in the tropical forests, which were left largely unscathed until recent centuries.

The human assault on the atmosphere probably started with the large-scale burning of fossil fuels, which began in northern latitudes

▲ ◀ *Hydrocarbon smog from traffic; smoke from a coal-fired power station. But there is more to pollution than meets the eye. More dangerous by far are the carbon dioxide (from both sources), the oxides of nitrogen from the cars, and the oxides of sulphur from the power station, that fall on distant countries.*

in the Middle Ages. The burning of fossil fuel does increase the atmospheric concentration of CO_2. After all, when coal is burnt, then CO_2 is released which was taken up by plants millions of years in the past. And when it is burnt in large amounts, then the CO_2 that was taken up in the ancient world over a period of, say, a million years, may be released into the modern world in just a few years.

In addition, many kinds of coal – including the kind found in Britain – contains appreciable amounts of sulphur; and when it is burnt, this is released as oxides of sulphur.

Ever since human beings first began to practise agriculture, 10 000 years ago, or at least since they learned to fertilize the fields, they have caused extra quantitites of nitrogen oxides and ammonia to flow into the atmosphere. This increased as agriculture grew more intensive, and as the number of livestock and the application of fertilizer has been increased. The invention in World War I of the Haber process – a method of making ammonia from atmospheric nitrogen – began the modern era of *artificial* fertilizers, and thus increased atmospheric pollution by compounds of nitrogen. The spread of the motor car, which pushes out oxides of nitrogen from its exhausts, adds to the problem.

Another serious side-product of agriculture is methane gas; belched not least from the mouths of the world's increasing population of cattle.

The world has invented many industrial processes which often create pollutants of a kind that do not occur in nature at all. Notorious

91

Pollution of the atmosphere – acid rain

The most immediately obvious of the present atmospheric threats is acid rain. Rain is naturally acid, with a pH* of around 5.6; as it falls it picks up carbon dioxide from the atmosphere, which dissolves to form carbonic acid. But in industrial countries in particular – or in countries that abut industrial countries – the rain has been growing steadily more acidic in recent decades, to the point where it has already done severe damage to ancient buildings and statues, where it is has virtually destroyed vast numbers of trees in continental Europe (notably Germany) and killed most of the life, including the fish and crayfish, in many of the lakes and rivers of Scandinavia.

At first sight, acid rain seems a simple enough phenomenon. European coal contains a small percentage of sulphur and when it is burnt, in power stations and factories, sulphur dioxide (SO_2) is produced, which (after further reactions with atmospheric oxygen) dissolves in rain water to produce sulphuric acid. Power stations and motor cars also produce various oxides of nitrogen (NO, NO_2, etc) which dissolve to form nitric acid. London in the 1950s suffered horrendous 'smogs' (smoky fogs) largely through burning coal in domestic grates; the one of 1952 had an estimated pH of 1.6,

* pH is a measure of acidity, extending from 1 (extremely acid) to 14 (extremely alkaline). pH 7 represents neutrality. The scale is logarithmic. That is, pH 2 is 10 times more acidic than pH 3, which in turn is 10 times more acidic than pH 4.

which made it more acid than lemon juice. Smoky fuels were then banned in most cities, North Sea gas came on tap, and factories built enormous chimneys (up to 300 metres high in the 1970s) to take the smoke away. But the coal was not cleaned up, and although British towns became cleaner, Britain continued to pour around five million tonnes of sulphur per year into the upper air, to be blown into Europe.

The effects of acid rain, however, are not simple. For example, it is not simply a drop in pH that has caused salmon and rainbow trout to disappear from thousands of Scandinavian rivers, and several large lakes. The lowering of pH in fact causes aluminium to be released from clays (in which it abounds), and this causes the fish's gills to produce mucus, which reduces their ability to absorb oxygen. Hence, the fish suffocate. Acidity can have more direct effects, however. For instance, it softens the shells of crayfish, and exposes them to disease. Acid rain affects trees partly because it dissolves essential nutrients out of the soil, such as magnesium and calcium – but again releases aluminium, which the trees take up to their detriment. Air pollution also apparently encourages the growth of fungi, and the spread of pests such as the bark beetle.

However, it is not clear precisely what should be done to ameliorate the situation. No serious thinker disputes that we must clean up factories and cars – removing the sulphur from coal, and the oxides of nitrogen from car exhausts. The technology exists

among these are the CFCs, chlorofluorocarbons, which are employed primarily in refrigeration. CFCs tend to inhibit the flow of infra-red radiation through the atmosphere; and they help to catalyze the breakdown of the ozone layer. So although CFCs are present in the atmosphere only in minute amounts, their effects are profound indeed.

These various gases – carbon dioxide, oxides of sulphur and nitrogen, ammonia, methane, and CFCs – pumped out from domestic fires, cars, industry and agriculture, are leading to atmospheric changes that have profound ecological consequences. These are the greenhouse effect, the destruction of the ozone layer, and acid rain. The first of these will probably lead to an increase in global temperature; the second could lead to an increase in the amount of ultraviolet radiation reaching the Earth's surface; and the third is already destroying vast areas of forest and lakes in northern Europe. These are described on pp.86–89, and above.

However, it is time to take a more positive view, and to look at the way early organisms began to combine energy from the sun with

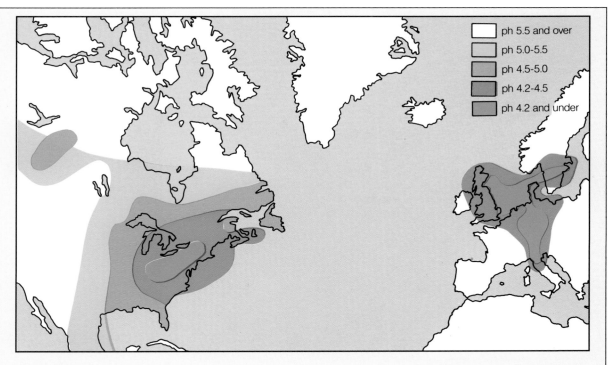

Legend:

- ph 5.5 and over
- ph 5.0-5.5
- ph 4.5-5.0
- ph 4.2-4.5
- ph 4.2 and under

▲ *Acidic gases can be blown long distances from their sites of release before falling as acid rain.*

for this. Liming would reduce the acidity of lakes, and forests could be helped in the short term by adding fertilizer. But the expense will be horrendous. Such programmes in Germany alone would cost an estimated $15 billion. Besides, there is a back-log of acidity already in the soil; and according to one estimate, Scottish lochs would remain at their disastrously acid level even if present acid fallout was halved by AD 2000.

We have benefited from our industry and our rapid transport. But the price is still to be paid.

gases in the atmosphere and in so doing enabled life as we know it to evolve.

Photosynthesis

The first living things to appear on Earth about 3.5 billion years ago probably resembled some of the more primitive bacteria that still exist today. They lived in water; perhaps in puddles, attached to clays. Wherever they were, they would have been bombarded with radiations; and as the primitive atmosphere was different in composition from that of today, and in particular would have lacked an ozone layer, they could well have received higher doses of radiation than they would today. Much of that radiation would have been potentially damaging: liable to break down the organic molecules of which those early organisms were composed.

Natural selection, then, could have favoured organisms that contained pigments whose molecules were able to absorb solar radiations, and then dispose of the energy, thus preventing

breakdown of more vital molecules. Some of those pigments presumably behaved in the way that melanin behaves in human skin, absorbing solar energy, disposing of it through the body as heat.

But other pigments in those ancient 'bacteria' developed other ways of disposing of the energy they had absorbed. It seems, for example, that more than three billion years ago, organisms evolved that could trap solar radiation within a pigment and then use the energy to split the molecules that existed in reasonable amounts in the atmosphere of very ancient times, such as hydrogen sulphide and ammonia. This splitting would simply have created more energy. But now this energy was in an easily useable form – chemical energy. And some of these early organisms then developed a further ability: to use that newly released chemical energy to manufacture their own organic molecules. These organisms, which could trap solar energy in a pigment, and then convert that solar energy into chemical energy, were self-perpetuating. They did not need to rely upon chance chemical events in the world outside to create the organic molecules from which they themselves were constructed. Through the intervention of radiation-trapping pigments, they were energized by the Sun.

It took hundreds of millions of years to evolve these abilities, longer, perhaps, than it has taken human beings to evolve from fish. Some of the organisms that evolved the ability to split hydrogen sulphide have left direct descendants, bacteria that now live in environments such as marshes and hot springs.

▶ ▲ *Plants do not have a monopoly on photosynthesis. Bacteria – like these modern cyanobacteria (above left), growing on the surface of the soil – evolved the ability first. Sulphur bacteria (above) practise a similar process, but split hydrogen sul; rather than water.*

But some of those chemical-splitting organisms evolved a further skill. Instead of splitting hydrogen sulphide, they developed the ability to split water by first trapping solar energy in green pigments. Hydrogen sulphide was probably a relatively rare commodity, even in ancient times. But water is everywhere.

The green-pigmented organisms that could split water inherited the Earth. The first of them resembled the still-existing cyanobacteria (formerly known – wrongly – as 'blue green algae'). Plants – at first single-celled, but later multi-cellular – inherited this ability from photosynthesizing bacteria.

The technique of splitting water to release chemical energy is half of the process known as *photosynthesis*. Each water molecule contains two atoms of hydrogen and one of oxygen – H_2O – so, as the molecules are split, oxygen gas is released. And as photosynthetic organisms spread throughout the world, oxygen increasingly became a major component of the atmosphere. There would be no free oxygen in the atmosphere even today were it not for photosynthesis. The destruction of the green plankton of the oceans, and of the tropical forests, would bring all animal life to an end.

Hydrogen is also released as water is split, and photosynthesis employs a second process to make use of it. The hydrogen is joined to carbon dioxide gas, CO_2, from the atmosphere, to make organic compounds with the basic formula, $(CH_2O)_x$. From these organic compounds, sugars or fats are created; and by the addition of nitrogen, and further chemical manipulations, proteins and nucleic acids are formed.

The leaves of plants are beautifully designed to carry out photosynthesis. They are flat, and usually are arranged either parallel to the ground, as in most trees, or vertically, like flags, as in grass: whichever arrangement intercepts the sunlight most efficiently in the particular circumstances.

Each leaf consists of a sandwich. Top and bottom are single layers of transparent, protective epidermal cells. Between these layers is a fairly loose, spongy arrangement of mesodermal cells. Chlorophyll, the green pigment that actually captures the Sun's energy, is contained in tiny 'organelles' known as chloroplasts; and these are mostly arranged within the mesoderm cells. The epidermis is penetrated at intervals by tiny pores known as stomata, which can be opened or closed by the surrounding guard cells, through which carbon dioxide enters (and oxygen exits). Water is brought into the leaf from the roots via the xylem vessels, which in the leaves themselves form part of the veins.

The main problem for the photosynthesizing plant is to keep the stomata open when the sun is shining, to allow carbon dioxide in without losing too much water by evaporation. Plants that grow in reasonably moist surroundings manage this easily enough, though they tend to wilt if their water supply is inadequate. Many specialist desert plants practise a peculiar form of photosynthesis known as *crassulacean acid metabolism* (CAM) whereby they open their stomata only at night, when water loss is liable to be at a minimum. They trap

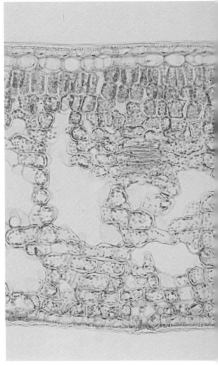

the carbon dioxide that enters at night in temporary chemical stores, and release it into the chloroplasts during the day, when the sun is shining again. But they keep their stomata firmly shut by day. Plants that practise CAM include the members of the (usually succulent) Crassulaceae (which is how CAM got its its name) plus many unrelated desert species, such as the cacti.

Note incidentally, that the term 'photosynthesis' applies to two different processes: first, the splitting of water; and second, the joining of released hydrogen to carbon dioxide, to create organic molecules. These two abilities presumably evolved separately; after all, the splitting of water is useful in its own right, because it releases some chemical energy. But the second process – the creation of organic molecules incorporating hydrogen – was the icing on the cake which gave the photosynthesizing organisms their supremacy. In existing plants and cyanobacteria, the two components of photosynthesis are very closely linked, and run as one smooth, continuous process.

In early times, then – certainly more than two billion years ago – photosynthesis evolved. From that time on, the photosynthesizing organisms emerged as the predominant life forms, and the atmosphere was altered by the increasing out-pouring of oxygen, until it reached its present state. We are the inheritors. However, the present chemistry of the atmosphere is a result of history. It is not an unalterable 'given', like the composition of the Earth's core. It is all too vulnerable to change.

▼ ▲ *Leaves are beautifully designed to capture the energy of light. They are arranged to intercept as much as possible. The cells that contain the chlorophyll that picks up the photons are mostly sandwiched in the spongy heart (the mesophyll) of the leaf. As shown in this cross section, these mesophyll cells are surrounded by air, which contains the necessary carbon dioxide. Desiccation is their main problem.*

Note, also, that oxygen gas – that fraction of the modern atmosphere that we and other animals find so essential – is a mere by-product; something that appeared in the atmosphere almost by chance, long after the first living things evolved. This is discussed in Chapter 1, p.12.

Our debt to the Sun

On a clear day on the equator, at noon, each square metre of the Earth's surface receives about one kilowatt (kW) of solar energy; that is, one hundred kilocalories (kcal) per hour. The amount reduces, of course, as we move away from the equator, and as we interrupt proceedings with nightfall and cloud. But even Britain receives an average of around 100 watts per square metre in each 24 hours; that is, an average of 100 kilocalories (100 kcal) per hour.

These figures have an abstract quality, and may not seem to mean much. Consider, however, that adult human beings need roughly 2400 kcals per day. This is the amount that beams on average on to each *square metre*, even in cloudy, wintry Britain. This means that if humans were able to absorb solar energy directly, with 100 per cent efficiency, then we could cram in one adult per square metre. In the USA there could be twice that density, and in the tropics (even taking night-time into account) there could be three people per square metre!

Fortunately, perhaps, life does not work like that. Before human beings (or any other animal) can make use of the energy from the Sun, that energy has first to be turned into chemical energy (that is, into complex organic molecules) by photosynthesis. In addition, the plant has to gather many other materials (water and minerals) in order to make proper use of the organic molecules it creates by photosynthesis. Photosynthesis does not operate at 100 per cent efficiency; and other vital ingredients (such as minerals or water) may be lacking. Thus wild plants are probably lucky to turn more than about one per cent of the solar energy that bathes them, into organic molecules; and even agricultural crops, bred for efficiency and well fed, are unlikely to get above five per cent.

Then again, no animal eats the whole of a plant – except in the case, say, of turnips, where it is possible to eat most of the root as well as the leaves. That proportion of the crop that can be eaten – or at least which is harvestable – is called the 'harvest index'; and even with well-bred crops such as wheat, the harvest index is unlikely to be above 60 per cent.

So in order to produce 2400 kcals of chemical energy in the form of organic molecules, you would need anything from about 40 to 100 or more square metres of land in Britain, and somewhat less than this in the USA. Even so, this is a surprisingly small area. It suggests that a human being could in theory obtain all the food he or she needed from a well-tended patio!

Such figures as these are really just for curiosity. They do show, though, that the world even now is not quite as crowded as it seems;

that if agriculture was better organized, then it probably would be possible to feed the present human population, and still leave plenty of room for the rest of nature.

The main point of this section, however, is that almost all the life on Earth that we are aware of derives its energy from the Sun, via the process of photosynthesis. Those organisms that do not derive their energy from this route (including some bacteria that derive energy directly from various tricks of inorganic chemistry) are now confined to obscure corners of the Earth (like the edges of undersea volcanoes) and play little part in the mainstream of Earth's ecology.

By far the majority of life on Earth, indeed, is underpinned by plants, which have evolved the trick of tapping solar power by photosynthesis; or at least, have borrowed that trick from bacteria. Animals in turn live by eating the organic materials that plants have produced by photosynthesis; or by eating other animals that in their turn have eaten plants. Thus we are all of us part of a food chain – or, rather, as chains are usually branched, of a food web – that is anchored, at bottom, in photosynthesis.

Food chains (or webs) are the subject of the next chapter.

▶ *In theory – even in cloudy Britain – the Sun provides enough energy to support a human being on every square metre of land. (In practice, life is not so simple).*

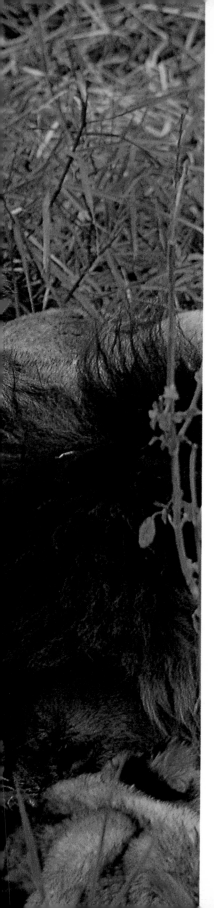

6 Food webs

The organisms that use the energy of the Sun to build organic molecules are known as *autotrophs*: Greek for 'self feeders'. Of the autotrophs, the most successful and widespread are the ones that split water, and attach the hydrogen thus released to carbon dioxide, in the process known as photosynthesis (see p.93). Plants do not have a monopoly on photosynthesis, and they did not 'invent' it. Bacteria did it first, and photosynthetic bacteria still exist, including the highly successful cyanobacteria. But plants are the most widespread and ecologically important photosynthesizers, and therefore the most important autotrophs.

There is one quite different, though common form of autotrophy. Many bacteria (including some cyanobacteria) are able to convert nitrogen gas from the atmosphere into ammonia, with the aid of pigments and enzymes. This process is known as *nitrogen fixation*. This is an extremely important source of nutrition for plants as they can take up the ammonia dissolved in water, or it can be converted to nitrates by other bacteria and then taken up. The bacteria that carry out nitrogen fixation are not necessarily complete autotrophs – although nitrogen-fixing cyanobacteria come very close to it. Most obtain the carbon that they need in organic form (see p.104).

Organisms that feed on organic material already created by some other organism are called *heterotrophs*, which is Greek for 'different feeders'. The heterotrophs include animals, fungi and most bacteria. Some green plants have also re-adopted some degree of heterotrophy and are parasitic or insectivorous. Evolution, above all, is opportunistic!

There are many ways of being a heterotroph. Animals either eat other animals, in which case they are called *carnivores*; or they are plant-eaters – *herbivores*; or (like ourselves) they eat a mixture of both and are called *omnivores*. Carnivores are sometimes referred to informally as *predators* – 'animals that prey'. Often, however – depending on context – biologists may also refer to herbivores as predators. After all, they prey upon plants.

To be an effective herbivore, or an effective carnivore, an animal needs special adaptations. These are described on pp.108–111. Omnivores, in general, require – and possess – a mixture of qualities. Most large animals at least are omnivorous to some extent. Deer or antelope commonly chew at bones, for example (and wild deer have been known to bite the heads off nesting birds, perhaps because they were calcium-deficient); and polar bears, the most carnivorous of bears, will eat kelp (seaweed) in the wild, and in captivity eagerly seize upon bark and herbs.

Fungi and bacteria commonly are *saprophytes*. That is, they tend to live inside or on the surface of organic material that is already dead, exude enzymes that digest that material, and then take in the digested nutrient, much as the roots of a plant take in nutrients from the soil. These saprophytes are the principal organisms of *decay*, or *decomposition*, and thus fulfil a vital role. They release all the nutrients in dead creatures, making them available to the next generation.

Many animals, fungi, bacteria, viruses, and some plants live as *parasites*. That is, they live on or within the tissues of other creatures without killing them immediately – or, indeed, without necessarily killing their hosts at all. Not all parasites even cause their hosts any great distress. But some do; and they, of course, are the agents of disease, otherwise known as *pathogens*. As described on p.132, pathogens are now known to be of enormous ecological importance.

For any creature, it is not enough simply to be adapted physiologically or anatomically to its particular kind of food. If animals are to succeed in feeding themselves, they must seek food – *forage* – efficiently. Accordingly, one of the flourishing subjects in ecology these days is that of *optimum foraging strategy*.

▲ *Decay is part of the food chain too. Saprophytic fungi and bacteria are its principal practitioners.*

◄ *Some plants have re-invented heterotrophy. Here, a sundew has captured and is slowly digesting a hoverfly.*

Optimum foraging strategy

In the course of a year (an annual cycle) every kind of animal has to consume enough food to keep it going, and perhaps to reproduce. If it does not, it starves or remains without offspring. Obvious.

Obtaining enough food, though, is rarely as simple as it seems. The point is not merely that food may be hard to come by. Equally important is for the animal to make the right *decisions*: what to eat; how much time to invest in eating each kind of thing; how much time to devote to eating, and how much to other things.

Consider, for example, an antelope, grazing. At first sight, its task seems easy. It has merely to consume until it is full, and then stop.

Well, if it was a sheep on a lowland farm, that probably would be all it had to do. On such a farm, after all, the grass would be specially bred for grazing, with appropriate levels of protein and energy, and should be in prime condition for eating – young and lush. In the wild, though, pasture is not like that. It contains many different species, each at various stages of growth; some young and tender, some old and rank. All grazing animals need to be selective, because they are equipped to eat only particular kinds of food: some geared to eat vast amounts of very fibrous material, others needing more nutritious, concentrated fare. In the wild, then, each grazing animal has in practice constantly to *select* what is best for it.

Suppose, though, that there is a shortage of the kinds of food that are best for the animal – as is very likely to be the case. Then it has to eat something second best which is more available. Then it has to make decisions. Should it spend a long time looking for the best food, which is rare, or make do on second-grade stuff which is easier to come by, but does not do it as much good?

There is urgency in such decisions. An antelope, after all, cannot spend all day, every day, grazing. It must spend some time at least, in the year, searching for a mate, or giving birth. It must spend some time on guard, or running away from predators. It must probably spend quite a lot of time looking for new grazing grounds, and perhaps for water.

In many cases the urgency is even more obvious. Some wading birds continue to feed on shellfish in estuaries and beaches even in winter in northern latitudes. But in northern latitudes the winter days are short, and the tides are out only for a part of each day. So the bird may have only a few hours to feed. If it made the wrong feeding choices – spent too much time seeking the best possible food, or ate too many shellfish that were insufficiently nutritious – it would starve.

Any animal that does survive, clearly has to some extent solved its own food problems. However, in order to survive, it must practise, as nearly as possible, an *optimum foraging strategy*.

At Oxford University, Professor John Krebs has made such strategies a special study. There are, he says, various ways in which an animal might in theory achieve the optimum. It might, for example, seek to maximize the speed of intake of food. Or it might

103

Nitrogen fixation: a form of autotrophy

Nitrogen is one of the essential components of proteins and nucleic acids, including DNA. Plants absorb nitrogen chiefly in the form of nitrates and ammonia; and they combine this with organic material made by photosynthesis. Animals derive the component parts of their proteins and nucleic acids ready-made from plants. In general, then, the supply of nitrates and ammonia for plants is one of the most pressing problems in all of nature.

Manures supply nitrogen combined in various organic compounds: but plants cannot usually absorb this nitrogen until those organic compounds are broken down (by bacteria) to release nitrates and ammonia. When farmers and ecologists say that a soil is 'high in nitrogen', they mean one that contains useful amounts of nitrogen-containing compounds – either nitrates and ammonia, or organic materials that will provide nitrates and ammonia when broken down. When they say that a soil is 'fertile', they usually mean that it is rich in such nitrogenous materials.

But when nitrogen is not combined with other materials to form compounds, when it simply exists on its own, as an element, it is a gas. In the form of gas, nitrogen is the principal component of the atmosphere, which contains about one thousand, million, million tonnes of it.

When it is in the form of gas, nitrogen is useless to plants. They cannot absorb it. However, many bacteria *are* able to capture nitrogen gas from the air, and convert it into ammonia. This process is known as *nitrogen fixation*.

The ammonia formed by fixation may then be converted (by other bacteria) into nitrates. Whether this second stage takes place or not, nitrogen fixation is one of the most significant sources of fertility both in the wild and in agriculture. It also provides the only natural source of *new* fertility. After all, the nitrates and ammonia obtained from the breakdown of manures merely represent a re-cycling of materials that were already present. Note, though, that farmers also add 'new' fertility partly in the form of inorganic, nitrogen-containing rocks, such as potassium nitrate, and partly through artificial fertilizers which are made by an industrial form of nitrogen fixation.

The bacteria that can fix nitrogen are of many different kinds, and live in many different habitats. Among the most important worldwide, are types that are able to carry out photosynthesis. These were for a long time known as 'blue-green algae' implying that they were plants, but they are not,
continued over

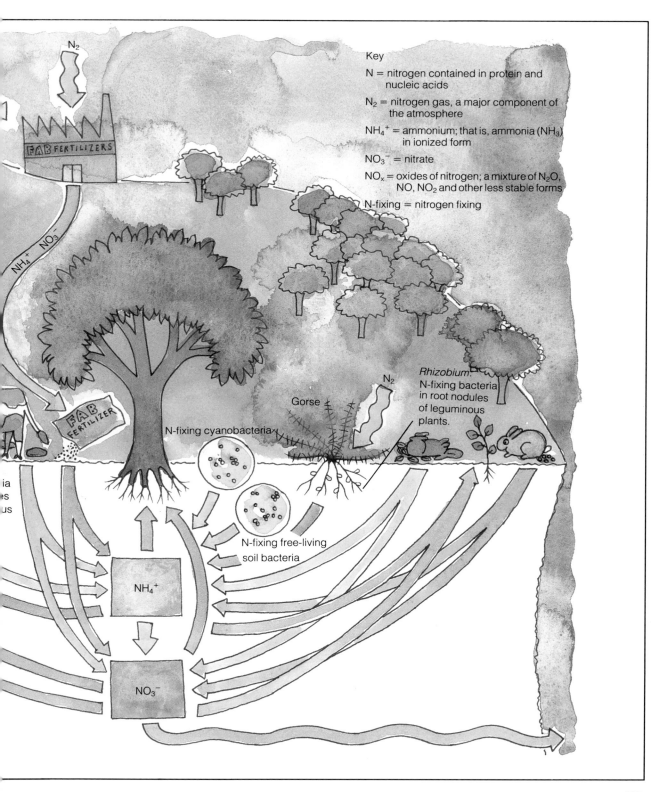

Key

N = nitrogen contained in protein and
   nucleic acids

N_2 = nitrogen gas, a major component of
   the atmosphere

NH_4^+ = ammonium; that is, ammonia (NH_3)
   in ionized form

NO_3^- = nitrate

NO_x = oxides of nitrogen; a mixture of N_2O,
   NO, NO_2 and other less stable forms

N-fixing = nitrogen fixing

N_2

FAB FERTILIZERS

NH_4^+ NO_3^-

FAB FERTILIZER

N-fixing cyanobacteria

Gorse

N_2

Rhizobium:
N-fixing bacteria
in root nodules
of leguminous
plants.

N-fixing free-living
soil bacteria

NH_4^+

NO_3^-

continued from previous page

and they are more properly known as *cyanobacteria* (meaning 'blue' bacteria).

Not all cyanobacteria are nitrogen-fixing. Some that are, are often seen lying on the surface of damp earth, forming a greenish scum. They are contributing to the fertility of the soil.

Even more intriguing are nitrogen-fixing cyanobacteria known as *Anabaena*, some of which live *inside* the leaves of tiny floating ferns known as *Azolla*. This is a symbiotic relationship, similar to the associations of algae with fungi to form lichens, or of algae with corals and clams as in coral reefs. Rafts of *Azolla* with their cargoes of *Anabaena* float in the paddy-fields of the Far East: and they are a very significant source of nutrient for the rice crop.

Most of the bacteria that fix nitrogen are not the photosynthesizing kind, however. Some of the ordinary nitrogen-fixing bacteria live free within the soil, where they again contribute significantly to fertility.

But other nitrogen fixers live within the roots of plants. This is yet another symbiotic relationship: the plant provides the bacteria with sugars, and the bacteria provide the plant with essential ammonia. Indeed, many plants have evolved special structures in their roots – nodules – to accommodate the bacteria. Plants that do this come from several different plant families; and they generally live in areas where soil nitrogen is low. The alder tree and the bog myrtle, for example, have nitrogen-fixing bacteria known as *Frankia* in nodules in their roots.

The plants that are best known for their symbiotic relationship with nitrogen-fixing bacteria are the legumes: clovers, vetches, lupins, beans and peas, gorse, broom, laburnum, acacia, and many others. The bacteria they carry are of the genus *Rhizobium*. Many ecosystems could hardly exist at all were it not for legumes and their associated rhizobia to provide fertility. For example wattles, which are acacias, are crucial to the Australian ecosystem. The semi-deserts of Africa rely upon acacia trees too, and one of the chief contributions that could be made to Ethiopia would be to improve the clovers on its hills.

We must add, though, that almost as fast as the nitrogen-fixing bacteria are creating ammonia and nitrates from nitrogen gas, other bacteria, of various types, are busily breaking it down again. Plants have to work quickly, to absorb their share. It clearly pays to do as the legume, the alder and the bog myrtle do – and keep the nitrogen-fixing bacteria on the premises.

seek not to consume as fast as possible, but to find food while expending minimum energy. The properties that the animal seeks to enhance – speed or efficiency – are known as the *currency* of the strategy, and part of Professor Krebs' research is to discover which currency each animal is using at any one time.

In theory, an animal could work out an optimum strategy by thinking about it; a leopard, for example, might work out whether two small rabbits that were easy to come by were worth one antelope that it had to spend time chasing. But biologists do not generally credit animals with such powers of calculation. Instead they assume that each animal has some simple set of rules programmed into it, that enable it to make appropriate decisions. For instance, the leopard might choose to catch rabbits if rabbits happened to be extremely common, and antelopes were rare; but would elect to chase antelopes if there were plenty about. Such a simple 'rule of thumb' could result in what appeared to be an elaborate strategy – and one that gave the desired result.

Study of optimum feeding strategy could have enormous theoretical value in helping us to understand why some animals succeed in some environments while others fail. It could also be of great practical value in conservation. For example, there is at present a proposal to build a barrage across the Severn River in the west of England, to extract tidal power. If we are to predict the effect of this

► *Optimum foraging strategy. Once, biologists assumed that animals simply took what they needed, as if the world was their cafeteria. Now it is clear that wild creatures survive only by making the correct – albeit unconscious – decisions. Always they have to weigh the small benefits of poor food that may be easy to come by, against the possible advantage of catching more difficult prey that is more nutritious.*

How to be a herbivore

Plant cell walls are made from a mixture of complex carbohydrates (and other, similar materials) of which the best known is *cellulose*. Cellulose is the most common complex organic molecule in nature, familiar in every day life in various forms; for example as cotton and as paper.

Cellulose consists of molecules of glucose joined end to end. In theory, then, it could be a tremendous source of food energy. But there is one huge snag. No animal of any kind is known to produce a *cellulase*: an enzyme capable of breaking cellulose down, and thus releasing its energy. It seems that nature has played the animal kingdom a very nasty trick; apparently making the world's most potentially important source of food energy unavailable.

Solutions: bacteria to the rescue

Some herbivorous animals – or at least those that derive a high proportion of their food from plants – overcome the energy problem by eating only those parts of the plant that contain food energy in forms that are readily digestible. Thus mice and many small birds eat only seeds, which are laden with energy-rich starch and usually with fat. In these, the energy theoretically obtainable from cellulose is of very little consequence.

But the animals that eat leaves and stems (apart from pandas) do derive most of their energy from cellulose. They do not produce a cellulase themselves, but have special guts that can accommodate vast armies of bacteria: and these bacteria *do* produce cellulases, which break down the cellulose. The relationship between the herbivore and its bacteria is one of symbiosis. The bacteria gain a safe and comfortable home, with a steady food supply; and the animal gains access to a huge source of energy.

The specializations of the gut to house these bacteria take various forms. In many herbivorous mammals, the bacteria live in a diversion of the hind gut, known as a *caecum* (pronounced see-kum). This is the case with horses, rhinos, and tapirs, elephants, rabbits and hares, specialist leaf-eating monkeys such as the proboscis and colobus monkeys, and koalas (which, of course, are marsupials).

But some of the most conspicuous and biggest herbivores contain a vast bacterial flora in their *fore-guts*; that is, in their stomachs, which are generally much modified for the purpose. Among placental mammals, the outstanding fore-gut digesters are the ruminants, which retain bacteria in a partition of the stomach called the *rumen*. These include cattle, antelope, sheep and goats, giraffe and deer, and the camels and llamas. Marsupials have their fore-gut digesters too – the kangaroos.

Among both fore-gut and hind-gut digesters, we find differences in overall feeding strategy. Some, such as llamas, red deer, and to some extent horses, eat fairly low-grade fodder which is very high in cellulose but is low in other nutrients. They consume it in vast amounts, but push it through their guts very quickly, extracting the best of what there is, but leaving a lot behind. Others, such as giraffes and many antelope, prefer richer fare: young leaves, rich in proteins and sugars, as well as fibre. Such food repays thorough (and prolonged) digestion. Cattle fall somewhere in between: they tend to like lush grass, neither too fibrous nor too delicate.

Most leaves, in nature, have evolved at least some methods of avoiding being eaten. Some are just too fibrous, or prickly, or toxic. Plants, though, like every other living thing, must balance the *cost* of whatever they do, against the benefit. It takes energy to create a spine, or a toxin, and it can be more cost-effective simply to endure being nibbled.

Many herbivores have developed specialized ways of overcoming the defences of plants. Perhaps the best example is the ability of the koala to overcome the numerous defences of eucalyptus, which is the commonest kind of native tree (with 600 species) in the dry lands of Australia. Wild koalas usually live exclusively on eucalyptus, compounding their diet

barrage on the dunlin and redshank, wading birds that now feed on adjacent mudflats, we must understand how they organize their feeding. If their chief food is wiped out, will they turn to something else? If they do, would they then be able to survive? What could be done to improve their food supply in ways that would truly bring them benefit?

There is much still to be found out. Overall, though, we see that different animals are primarily adapted to exploit different from around six different species. The leaves of eucalyptus are extremely fibrous, are high in tannins (which inhibit digestion of protein, and are toxic in large amounts), and are rich in essential (aromatic) oils which tend to be toxic. But koalas have specialist livers, able to detoxify the leaves and their caeca can separate the digestible parts of the leaves from the genuinely non-digestible, and push out the latter while retaining the former for their gut bacteria to work upon.

Not all herbivores have acquired the necessary bacteria, however, or evolved the guts to harbour them in. A notable failure is the giant panda which is closely related to the bears. Bears began as meat-eaters, but went on to become omnivores – eating vegetation as well as meat. But the panda took things to extremes, and now subsists mainly on bamboo. Yet it has never acquired the trick of digesting cellulose. Its bamboo diet provides it with adequate protein but very little energy. Accordingly, the ecology of the panda is geared to doing as little work as possible – although, because it has no spare energy to accumulate fat in summer, the panda is the only bear (if you call it a bear) that cannot hibernate.

But animals that can utilize cellulose also have a problem. Cellulose-rich food is extremely bulky, relative to its energy content, and digestion by cellulase-producing bacteria is relatively slow. But warm-blooded animals – mammals and birds – require a lot of energy. Large warm-blooded animals, however, require less energy relative to their weight than do small animals, because they conserve their body heat so much more efficiently. So warm-blooded herbivores that rely on cellulose for energy tend to be the large ones; bigger than a rabbit among mammals, or a grouse (which eats heather leaves) among birds. The smaller warm-blooded herbivores are the seed-eaters.

'Cold blooded' animals need far less energy weight for weight than warm-blooded ones; and specialist insects, for example, also employ gut bacteria to digest cellulose. The most notable examples are termites.

▲ *Eucalyptus leaves are appalling food: high in fibre, low in nutrients, and extremely toxic. But there are enormous amounts of them in their native Australia, and the pickings are rich for the few creatures that can cope. The koala copes best of all.*

components of the environment, and that each one practises the particular strategy that enables it to get by. Between them, living things efficiently carve up the resources of the world. In general, the autotrophs – and especially plants, in the modern world – start the whole thing rolling by tapping the energy of the Sun; and the heterotrophs consume the autotrophs, or each other. We should explore this general pattern in more detail.

How to be a carnivore

Most animal flesh provides much richer fare than most vegetation. Beef, for example, with the fat left on, provides more than 250 kcals of energy per 100 grams, plus 16 grams of protein; while cabbage offers only about 25 kcals per 100 grams, and less than two grams of protein. Seeds are a different matter: rice gives more than 350 kcals per 100 grams, and around 6.5 grams of protein. But seeds – in nature – are not common, and tend to be seasonal. Only small specialists such as mice and weevils can afford to live on them. Humans are the only large animals to subsist on seeds, by virtue of agriculture.

Carnivores, with such rich diets, tend to have short guts, lacking the vast side-passages that are usual in herbivores. They produce powerful enzymes that may operate in extremely acid conditions (low pH). They also commonly have livers able to deal with toxins (as might be produced by bacteria in rotting carcasses).

For those that can cope with it, animal flesh clearly is excellent food. Yet there are snags. First, the prey animals are generally better able to take care of themselves than are plants. Attempt to eat them, and they run away, or fight back. All carnivores tend to be good fighters, with powerful jaws, stabbing or gripping teeth, sometimes with dangerous claws, and with excellent senses and reflexes. Some also co-operate in hunting. Thus leopards, which do not generally hunt co-operatively, have to make do on fare that is smaller than themselves, such as antelope or monkeys. But lions, wild dogs and wolves, which do co-operate, can hunt animals much bigger than themselves: moose and wapiti, zebra and buffalo.

Carnivores must do more than defeat their prey. They have to be extremely good strategists. They have to 'know' which animals they are capable of killing, and which are liable to do them harm, and

then to balance risks against rewards. Thus a cheetah could dine out for a week on a zebra (provided it was not first chased off by lions or hyaenas), but it would be ill-advised to try to kill one. Even lions sometimes get their teeth kicked in. A lion, on the other hand, might find a lizard easy fare, if one happened to be passing. But a lion that ate lizards alone would soon starve. Wolves, however, in the Arctic summer, may well subsist for long periods on such humble fare as mice. At such times they may be so abundant that it is cost-effective to seek them out.

The skills required to catch prey, which may include co-operation and the ability to operate strategically, imply high brain power; and indeed, the specialist mammalian carnivores are among the most intelligent of animals. We should not get carried away, however. Many herbivores or omnivores are also extremely bright, including elephants, rats, and monkeys. And carnivorous birds such as hawks – again contrary to myth – are not particularly bright; certainly not compared with, say, herbivorous or omnivorous parrots or crows.

Animal flesh that is most worth having tends to come in large packages: an entire gazelle, for example. The safest place to keep spare food is inside one's body; so another worthwhile adaptation is the ability to consume vast amounts fairly quickly. A lion can easily consume 20 per cent of its own weight at a sitting: equivalent to a man eating 12.5 kg or more of steak, at the prospect of which even a Texan would baulk. After such vast intakes, there is nothing to do but wait until hunger descends again; big cats in particular are stupendously inactive. A wild lion typically dozes or sleeps for at least 20 hours a day.

Large carnivores also tend to be opportunist. In general they are not such specialists as the herbivores; they hunt whatever is available. Neither

Food chains, food webs, and cycles

In practice, autotrophs support not only themselves but, ultimately, all the heterotrophs as well. A very common pattern is for a plant (such as grass) to grow, courtesy of the energy from the Sun and the inorganic nutrients and water in the soil; the plant is then eaten by a herbivore – say, a hare; and the herbivore is then in turn eaten by a *primary predator* – say, a fox. The fox might in its turn be eaten by an even bigger carnivore, which thus becomes a *secondary predator*; for example, a lynx (although predation of carnivores on carnivores is uncommon amongst land mammals, probably because other

are they 'proud' as the old romantic naturalists liked to suppose. It simply is not the case that lions eat only what they kill, and that hyaenas come later to crunch the bones. Equally often, things work the other way round. Hyaenas are accomplished killers, and lions have an entirely professional attitude to hunting. They do not do it if they do not need to. Note, finally, that the 'professionalism' of specialist carnivores extends to their aggression. Carnivores are no more 'aggressive' than any other wild creature, unless they have specifically geared themselves up for a kill, in which case they need to get the adrenalin running. If you had to choose between being locked in a pen with a wolf or with a wild bull (or a boar or a stallion), then you would do well to choose the wolf. In the wild, wolves (and lions and leopards) take pains to avoid humans - again because they are good strategists, and attacks

▲ *Romantic naturalists liked to believe that lions hunt and hyaenas scavenge. In practice, the roles are just as likely to be reversed.*

on humans bring retribution. Big, vicious, domestic dogs, which people tend to think of as typical carnivores, are an aberration. They are *bred* artificially to be ill-natured, and often (partly because of in-breeding) lack the intelligence of their wild ancestors. If they displayed such uncritical aggression in the wild, they would quickly fall foul of some unsuitable prey, or else die because they expended too much energy for no purpose. In short, big, ill-trained modern 'fighting' dogs, such as Rottweilers or pit bull terriers, are far *more* dangerous than wolves or leopards. It is odd that we have declared war on the wild animals, but allow the 'domestic' ones to live in our cities.

carnivores can escape easily, and are relatively rare (see p.115)). Such a sequence is known as a *food chain*. The organism at the bottom of the food chain is always an autotroph (which usually means a plant), and the creature at the top of each chain, who is eaten by no other, is the 'top predator'. Each stage in the chain is known as a *trophic level*.

In some chains there are no 'top predators'. No carnivore regularly eats elephants, for example (though the bigger sabre-tooth cats, now extinct, may well have done); and no predator habitually eats large whales, though sharks may take a small toll. So the food chains that run, 'plants – elephants' , or 'plankton – krill – blue whale' are very short indeed.

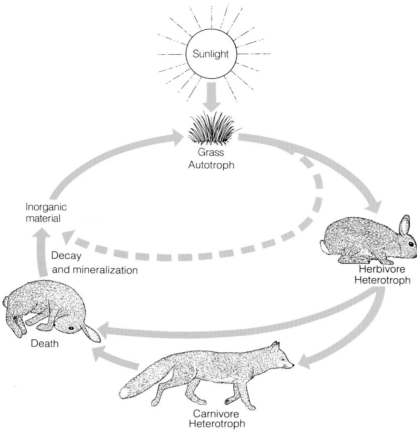

Sunlight

Grass
Autotroph

Inorganic
material

Decay
and mineralization

Death

Carnivore
Heterotroph

Herbivore
Heterotroph

But most food chains are nothing like as simple as this. Sometimes stages in the chain are missed out (eg lynxes are more likely to eat hares than foxes). Sometimes the top predator may be an omnivore, and also feed directly upon vegetation – as most kinds of bears are likely to do. Usually, any one predator may feed on several carnivores, and several herbivore species; and any one herbivore (or small carnivore) may be preyed upon by several different carnivores. Thus the sequence, usually, is not so much a food chain, as a *food web*.

If any one component of a food chain is disrupted, then all the other creatures that are part of it must adjust. Some will suffer, though others may flourish, at least temporarily. Again, the oceans provide clear-cut examples. For example, the krill which blue whales feed upon in Antarctica, are also principal or important food both for crabeater and for leopard seals. Leopard seals also attack and feed upon young crabeaters. So the reduction of the blue whale (and other krill-feeding whales) has increased the food supply for both the crabeaters and the leopards – the leopards benefiting both from the increase in krill and from the flourishing of crabeaters. Both species have burgeoned, then, as the whales have diminished. Crabeaters are the commonest seals on Earth: one estimate in the 1970s put the population as high as 50 million. The demise of the whales is a tragedy. But it's an ill wind ...

▶ *High latitude habitats have few species, and the food webs are accordingly simplified. This is the economy of the Antarctic. When baleen whales are depleted – by human hunters – krill increase; and so do the crabeater seals, which eat krill, and the leopard seals which prey both on krill and on crabeaters.*

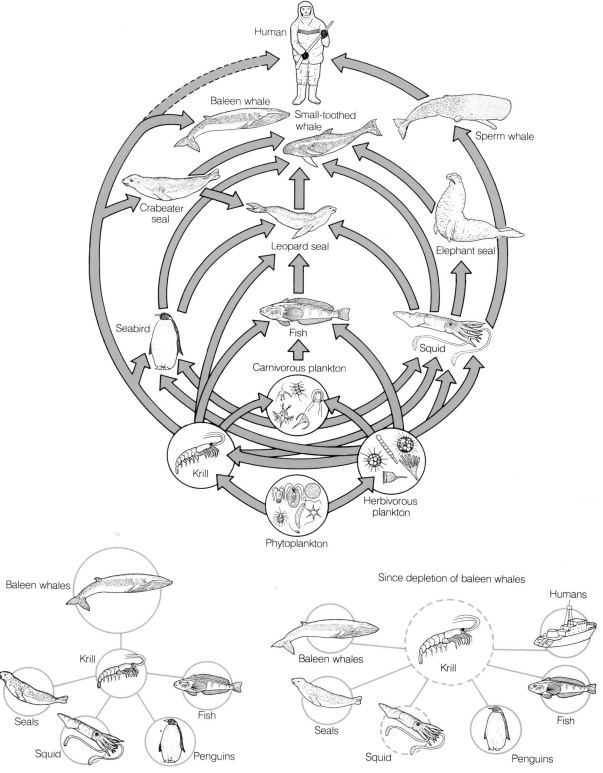

Human

Baleen whale

Small-toothed whale

Sperm whale

Crabeater seal

Leopard seal

Elephant seal

Seabird

Fish

Squid

Carnivorous plankton

Krill

Herbivorous plankton

Phytoplankton

Baleen whales

Krill

Seals

Squid

Fish

Penguins

Since depletion of baleen whales

Humans

Baleen whales

Krill

Seals

Squid

Fish

Penguins

Whales, however, are top of their food chain. If a component is removed from the *base* of the food chain (or web), then *all* the creatures above it are affected. Thus puffins in the North Sea suffer because fishermen have removed so many herring; and the consequences of a major oil-spill, killing all the planktonic creatures, are horrendous.

The top predator (or the elephant or blue whale) must die, however; and many animals or plants lower in the food chain also die of 'old age', or from disease, without being first consumed by something else. When any creature dies, the process of breaking down begins. Dead animals may first be torn apart by large carnivorous scavengers. As the corpse becomes smaller, and the pickings leaner, smaller creatures take over: vultures or marabou storks on land; crabs in the sea; and surprising passers-by such as porcupines or squirrels (which love to chew bones). Ants, flies, and beetles take their toll. Worms, protozoa, and fungi may play a part. But the final breakdown – *decay* – is effected, invariably, by bacteria. Note that bacteria also break down the waste organic materials that are excreted by all the animals that have fed upon the corpse; and break down their bodies (and those of the fungi, and the rest) when they die; and are in turn broken down, when they die, by other bacteria.

Final decay means that the carbohydrates and fats are broken down to carbon dioxide, which floats off into the atmosphere, and water; while the nitrogen in protein and nucleic acids is released in inorganic form, as ammonia or as nitrates. Plants can then take up these inorganic nitrogenous compounds through their roots, and the carbon dioxide through their leaves. Thus the *cycle* is completed.

Energy and trophic levels

There is a tremendous loss of energy between one trophic level and the one above. For one thing, every kind of animal needs to consume several kcals-worth of energy in order to build one kcals-worth of flesh. In the 1950s and 60s, biologists tried to put a figure on this, and suggested that the loss was about ten-fold. But the reality is far more complex. 'Cold-blooded' animals (such as fish, reptiles, and insects) may well be able to subsist on food resources that are able to provide only a few times as much energy as they themselves eventually contain. But warm-blooded animals (mammals and birds) use enormous amounts of energy in creating heat, and they may need hundreds of kcals for every one kcal that they themselves contain. Thus the biomass of grass must outstrip the biomass of zebras hundreds of times; and that of zebras must outstrip the lions that feed upon them hundreds of times.

This is not necessarily easy to see, because what really counts is not the mass of organic material that is visible at any one time (the *biomass*), but the turnover, the amount of organic material that is produced within a particular trophic layer within a given time. Thus, as we have seen, parts of the ocean produce an enormous amount of

▲ *Trophic levels. It takes an enormous amount of vegetation to support one herbivore; and a lot of herbivores to support one predator. So big predators need vast territories. Inevitably, then, the big cats, wild dogs, and bears are few in number; and so they are easily driven to extinction.*

plant material in a year in the form of algae, even though the *amount of algae* present at any one time, may be low. But the generalization does hold; the world can support far fewer tigers than it can deer and antelope, on which those tigers feed. In addition, animals in general are at least as big and often far bigger than the animals on which they prey.

Put both these facts together, and we see why it is that animals such as tigers are bound to be thin on the ground; and if you take a safari in Africa, you will see thousands of antelope and zebra, but only a few lions and leopards. It follows, though, that predatory animals are extremely vulnerable in the modern world: there are so few of them to begin with, that they are all too easily wiped out. We also see why there are few large predators on islands. There simply is not enough prey for them. Predators such as imported dogs and cats do well on islands because they are at least in part domestic, and human beings subsidize their diet. Mongooses do well in the Caribbean because they feed largely on rats, which in turn feed upon crops. Thus, mongooses in the Caribbean are part of a food web that is being enriched artificially by human beings, who bring in fertilizers from outside and thus augment the lowest trophic layers, which support the rest.

Why, though, aren't wild ecosystems more productive than they are? Why can't the output of farming be raised indefinitely?

The answers are many and complex, and ultimately lie in the innate abilities of biological systems to respond to inputs, and to grow. In practice, however, growth in the wild – and in farming – is constrainted by *limiting factors*.

Limiting factors

Plants need sunlight, warmth, water, carbon dioxide, and other nutrients. The only one of these whose presence can be guaranteed is carbon dioxide – and even that may be present in less than ideal amounts, on a hot, moist day, when plants are growing well. Certainly many greenhouse plants benefit from *added* carbon dioxide. Below the top level of the oceans, sunlight is lacking. In high latitudes, both sunlight and warmth are often deficient. In deserts, there is liable to be too little water. In most wild environments, there is far less available nitrogen than plants can theoretically make use of, and often a relative deficiency of phosphorus or some other nutrient. Chains are as strong as their weakest links; and if any one essential input is deficient, then plant growth will be geared to the one that is least abundant, and the input that is most deficient is therefore called the *limiting factor*.

Of course plants do adapt to adversity, for example desert plants make use of every milligram of moisture. But overall productivity is bound to be compromised in the places where such adaptations are actually necessary.

However, many ecosystems are fed by nutrients that come from elsewhere; inputs known as *subsidies*.

Subsidies

Mountain tops are 'subsidized' to some extent by the corpses of insects blown in from the lowlands. Ponds and lakes gain nutrients washed in from the surrounding hills. Estuaries are among the most enriched environments of all, with silt and nutrients of all kinds – organic and inorganic – washed down from the rivers, and deposited where the river meets the sea. Thus the silty mud of estuaries is rich in detritus feeders such as clams and mussels – which in turn, if unpolluted, support vast armies of wading birds. Indeed, many estuaries are among the world's most important habitats for wildlife.

Each individual animal in an enriched system such as an estuary can consider itself subsidized as well. Mussels, for example, feed by filtering out organic particles from passing currents of water. They do not need actively to pursue their prey, as most animals do. Their method of feeding is effective, however, only because the water is doing their work for them; bringing their food to them. Thus the current is effectively providing an *energy subsidy*.

However, although the growth of plants tends to be restrained by some limiting factor, and although many creatures living in a state of nature gain from subsidies from elsewhere, most wild plants are not able to cope with *too much* nutrition. Thus, if wild plants are given more nitrogen fertilizer than they are adapted to make use of, then most species tend (if they respond at all) to 'outgrow their strength'. Typically, they become long and spindly, fall over, and quickly succumb to disease. In general, the plants that do succeed in highly fertile soil are the ones that tend to flourish on farms, where they are

▲ *Estuaries worldwide are foci of life. Wading birds are the most visible; but they are feeding upon molluscs, worms, and crustacea, ensconced in the mud beneath. These invertebrates in turn feed upon organic detritus, brought down the river and deposited where the river enters the sea. Thus are estuaries subsidized, in nutrient and energy.*

categorized as 'weeds', while crops, of course, are in general *bred* for their ability to respond to high doses of nitrogen by producing more of whatever it is that the farmer requires. It may seem ironic, though, that fertilizer can be a serious pollutant, for example in the remnants of wild ground that serve as reserves among the arable fields of Western Australia. Excess nitrogen encroaching from outside upsets most of the native, wild species, while encouraging weeds to grow. Near cities, the faeces of dogs taken for walks among remaining patches of wildlife, can raise the nitrogen levels enough to change the entire balance of the native flora.

Ponds and lakes worldwide are particularly vulnerable to over-nutrition because they act as dumps for nutrients from far and wide. Excess phosphorus in particular can cause dense 'blooms' of algae and cyanobacteria. As the excess plant life dies, its decay mops up all spare oxygen, and everything else dies. This form of pollution – beginning ironically with super-abundant growth – is called *eutrophication*.

Some areas, however, *lose* nutrients to others. The surface layers of the sea are the great providers for all that lives below, as the phytoplankton photosynthesizes, feeds the zooplankton, and then both are eaten, or die and fall to the bottom. Were it not for the winter storms, many surface areas of ocean would remain permanently depleted.

The ideas that relate to the principles of limiting factors and subsidies help us to appreciate the true significance of agriculture; the greatest promoter of ecological change in the modern world.

Agriculture and the rise of human beings

As we have seen, the growth of all wild plants – and hence the status of all the animals that feed on them – is ultimately constrained by limiting factors. The central aim of agriculture and horticulture is to remove those limiting factors: to supply water, when it is needed, by irrigation; to furnish extra nutrients in the form of fertilizer; to provide extra warmth, if necessary, with greenhouses; and even (in modern intensive horticulture) to provide extra carbon dioxide.

The limiting factors are removed through subsidies: by taking materials from elsewhere. Organic farmers may gather manures from far and wide to increase the fertility of their fields. Modern 'conventional' farmers add nitrogen fertilizer that has been manufactured by converting nitrogen gas from the atmosphere into ammonia and oxides of nitrogen – processes that employ large amounts of energy from fossil fuel or nuclear power.

Wild plants also, of course, face competition from other species, and from their own kind. Farmers and growers seek to eliminate this competition; destroying wild competitors (weeds); killing parasites and predators (diseases and pests) and spacing the crops to reduce competition between plants of the same kind.

Modern crops are bred to respond *positively* to extra inputs. Wild wheats grow tall and fall over (lodge) if given more than a tiny amount of nitrogen, but modern semi-dwarf varieties stay short, and produce more grain. Modern potatoes, well fed, produce massive crops of tubers. And so on.

Modern crops, then, bred and cosseted, can and often do out-produce most wild ecosystems many times over. Hunter-gatherers living in a state of nature may need several square kilometres to feed themselves, but intensive farming communities could get by with a fraction of a hectare per person. Human beings have been practising farming, over ever greater areas and with ever increasing intensity, for at least 10 000 years; and it is because of farming, and only because of farming, that our population has been able to grow as vast as it has. After all, when farming first began, the total world population of humans was probably not more than around 10 million. Now it has reached 5 billion – 500 times as great; and it may peak at 10 billion (a 1000 fold increase) before it begins to decline again. In short, solely because of agriculture human beings seem temporarily to have broken the ecological law which says that 'big, fierce animals are rare'. If we did not increase the fertility of fields as much as we do, and protect our crops as assiduously as we do, then we would be at least as thin on the ground as chimpanzees.

A great deal of modern farming, however (not all of it, but certainly a proportion) extracts more from the soil than it puts back. Some of the inputs that are expropriated for agriculture from non-farming areas, such as phosphorus, could begin to run out – at least in easily accessible form. In vast areas, the soil itself is disappearing through wind and water erosion, organic material near the surface is being

▲ *The growth of plants in the wild, and therefore of animals, is always restrained by 'limiting factors': a lack of nitrogen, water, or whatever. In intensive agriculture – and especially in intensive horticulture, shown here – all the deficiencies are made good. Thus, intensive cultivation produces far greater yields than is usual in the wilderness; and thus farming can support an enormous (but precarious) population of humans.*

oxidized away, desert continues to spread, and more land is currently becoming saline than is being restored. It has been said (notably by the American ecologist Paul Ehrlich) that modern farming to a significant extent is *mining* the soil: causing permanent depletion and degradation.

If this degradation cannot be halted, then present outputs cannot be sustained. If they cannot, then famine must result – and as human beings become more desperate, they are liable to do even more damage than at present to the rest of nature. We can see our present state as a race. Can we devise systems of agriculture that are truly sustainable, but which nevertheless produce enough food for the truly prodigious numbers of people that are liable to inhabit this Earth in a few decades' time? Alternatively, will world populations reduce to the kinds of levels that could be maintained by sustainable outputs, before famine overtakes them? The next few decades will tell.

For now, we should look more closely at the biology of populations, and at the ways in which organisms come to terms with members of their own kind, and with others that share their habitat.

This is the subject of the next chapter.

7 Populations and communities

A group of creatures of the same *species* (see p. 124), living in the same area and so able to interact with each other, is called a *population*. A group of creatures who may or may not be of the same species, and simply happen to be in the same place at the same time though without necessarily interacting, is an *assemblage*. A group of various species – animals, plants, fungi and bacteria – all living in the same place and carving up the habitat between them, is called a *community*. All living creatures have to come to terms with their physical environment – temperature, moisture, light and dark and so on. The study of how they do so is half of ecology. But all living creatures also have to come to terms with others of their own kind, and with those of a different kind. The study of populations and of communities, then, is the other half of ecology.

The nature of populations

It is really only in recent decades that biologists have come fully to realize just how complicated populations really are: how many different forms a population may take; how many factors determine population size; and – increasingly – how *vulnerable* populations may be.

Some plants (and a few animals) have abandoned sex altogether and breed only asexually, and many other plants (and quite a few animals) breed both sexually and asexually. Among plants, some are natural 'in-breeders', and are happy to practise incest, so that they may fertilize their ova with pollen from the same plant (or indeed from the same flower); others, the out-breeders, *must* take in pollen from an individual that is genetically dissimilar to themselves (even though it must of course be of the same species). What consequences do these observations have on the meaning of 'population'?

Well, a patch of grass is a population – provided all the individual plants within it are of the same species, which is not always the case. If the grass has reproduced itself by seed, and if the grass – like maize – is a natural out-breeder, then all the individuals within that population will be genetically dissimilar, even though they are all of the same species. But a grass such as *Festuca ovina*, the sheep's fescue,

121

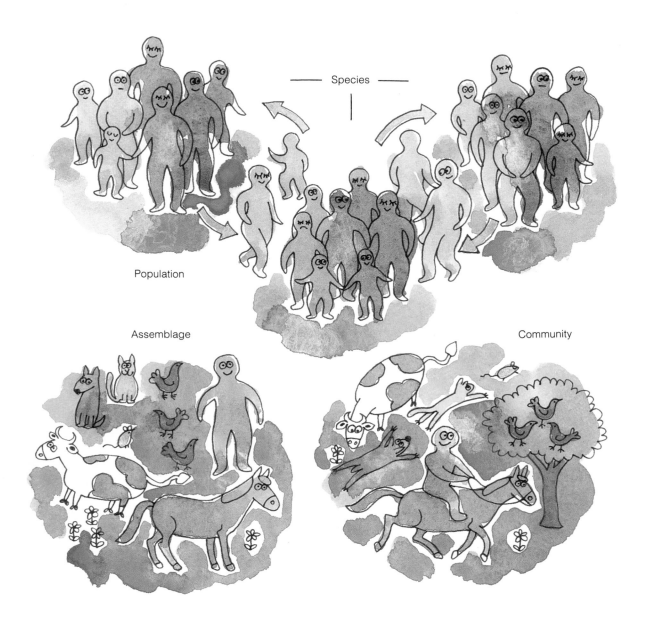

Species

Population

Assemblage

Community

▲ *Species (see p.124) of animals or plants may or may not be divided into several different populations – which may or may not be in regular contact with each other. If animals (or plants) of the same kind or of different kinds just happen to be in the same place at the same time, then they are said to constitute an assemblage.*

Often in nature, however, different creatures that are in the same place at the same time react to each other, and may even depend upon each other. When such relationships can be perceived, the assemblage can truly be called a community.

which is a common field grass in Britain, also reproduces simply by producing new stems (tillers); and a large patch of *F. ovina*, which may resemble a population, may, in fact, be a single plant. Indeed, it is thought that some of the sheep's fescue in some ancient British pastures are more than 1000 years old.

On the other hand, plants such as the dandelion, *Taraxacum officinale*, reproduce by seed – but the seed is unfertilized. In other words, they practise *parthenogenesis*. Thus in any one group of dandelions all the individuals may be genetically identical one to another even though each is clearly physically separated from all the others. Such a genetically identical group is called a *clone*. Grasses such as wheat do not practise parthenogenesis, but they are natural inbreeders. They may become so inbred, indeed, that patches of wheat plants may be similar enough genetically to resemble a clone – but in fact they are not, because each individual, though similar to its siblings, is the product of true sexual reproduction. In contrast to both of these are plants such as plums, which are so averse to incest, that they cannot breed at all unless the two parents are of different varieties (even though they are of the same species). From these examples we can see that the word population, in the real, wild world, covers a multitude of entities. Animals, fortunately, are not in general so variable as plants in this respect, and most of the conspicuous types practice conventional sexual reproduction. Some, though, such as the aphid (greenflies and blackflies) alternate sexual reproduction with several generations of parthenogenesis, and a

▶ *Aphids practise both sexual and asexual reproduction. Their sexuality ensures variation within the population, which helps them to cope with disease. But asexual reproduction can be more rapid, and enables them to colonize new host plants quickly. These are black bean aphids on broad beans.*

The meaning of species

It was the Swedish biologist Carl Linnaeus who, in the eighteenth century, first introduced the modern system of classifying living things. He did not believe in evolution; like most people before Charles Darwin, he believed that God had created all creatures individually, and more or less in their present form. But his system of classification was none the less 'natural' in that it does generally show the true relationships between animals and plants. That is, Linneus considered collections of animals or plants to be in the same group if they had what he felt were fundamental qualities in common, such as the same number of legs, or the same basic form of flower. Nowadays (after Darwin), we appreciate that when particular groups do have such qualities in common, it is because, at some point, they have shared a common ancestor.

Linnaeus called each 'kind' of creature a *species*. But he also grouped similar species into larger groups, called *genera* (singular *genus*); and grouped similar genera into *classes*, and classes into *kingdoms*. Each species was given two names - the name of its genus (generic name) plus its own 'specific' name.

Thus a domestic dog is called *Canis familiaris*. Other members of the genus *Canis* include the grey wolf *Canis lupus* and the coyote *Canis latrans*.

Since Linnaeus, further categories have been added. Genera are grouped into *families*, families into *orders*, orders into *classes*, classes into *phyla* (singular, *phylum*) and phyla into *kingdoms*.

All the members of the genus *Canis* are grouped, together with various genera of fox (including *Vulpes*) and the African wild dog (*Lycaon pictus*) and some others, in the family Canidae. The Canidae are members of the order Carnivora (together with cats, weasels, racoons, civets, bears, seals, sealions and walruses) in the class Mammalia. Mammalia are members of the phylum Chordata, in the Kingdom Animalia, or animals. Each kind of grouping is called a *taxon* and each taxon can be further subdivided (eg into subfamilies, or sub-orders) or grouped (eg into superorders).

Most of these groups are arbitrary, in the sense that different biologists group the various species in different ways, according to different criteria. For example, human beings are generally placed in the genus *Homo* in the family Hominidae, while chimpanzees (*Pan*), gorillas (*Gorilla*), and orang utans (*Pongo*) are usually put in the family Pongidae. Modern evidence suggests, however, that humans and chimps are extremely closely related, and are closer to each other than either is to the gorilla, while all three are distant from the orang. It would be more 'natural', then, to place chimps, gorillas, and humans in the same family (Hominidae); and even to place chimps in the genus *Homo*. This kind of issue is more or less unresolvable; it depends whether you think that true genetic relationships are most important, or whether other attributes should be considered as well. Humans, after all (some argue), have a quite different *culture* from other animals; and this alone (some feel) should place them in a separate category.

But the concept of species is more solidly founded. Definitions vary, but most follow the lines that 'two creatures may be considered to be of the same species if they can breed together sexually to produce fully viable offspring' – assuming, that is, that they are of the appropriate age and sex, and are living in the same place so that they can actually mate.

Such a definition solves many conceptual problems. Domestic dogs and domestic cats are clearly of different species. Even if they were to mate, they would not produce offspring. Horses and donkeys are different species, because although their union does produce offspring, those offspring – mules – are sexually infertile. Thus the offspring are not 'fully viable', and the species of donkey and horse shall forever remain separate. Dobermann pinschers and bulldogs are of the same species, though; they are merely different breeds of *Canis familiaris*. They may look very different, but if they mate, they can and do produce perfectly viable and sexually fertile puppies.

No single definition can ever cover all the eventualities of nature, however. Hooded crows live in Scotland, and carrion crows in England. The two species can and do interbreed where they meet, and the offspring seem perfectly fertile and in no way inferior. Yet the hybrid offspring do not spread into hooded crow territory, or into carrion crow territory. So although, according to reproductive biology, hooded and carrion crows should be of the same species, in practice they remain separate; and presumably they know their own business best. A similar band of non-encroaching but apparently viable hybrids marks the border between yellow-bellied toads and fire-bellied toads across mainland Europe.

Then again, some organisms have lost the ability to breed sexually. Such organisms include many that superficially look as if they ought to be able to breed sexually – including the dandelion, *Taraxacum officinale*. If two creatures cannot breed together sexually, then, according to our definition, they cannot be of the same species. Indeed we would be forced to conclude (if we wanted to be entirely purist) that each individual dandelion belonged in its own species. Yet that would be absurd. In practice we all know what a dandelion is, and are happy to acknowledge the specific status of dandelions. Rigid definitions must always be tempered by common sense.

Often, too, we do not *know* whether two creatures belong to the same species, simply because they do not have the *opportunity* to breed in nature – they may live in different patches of the same tropical forest, for example. This could in theory be tested, but there are too many animals and plants to put them all to the test!

So species does refer to a 'real' grouping; but even here, the richness of nature outsmarts our attempts to impose a simple view.

◀ ▶ ▼ *The grey wolf,* Canis lupus, *and the domestic dog,* Canis familiaris, *belong to the same genus –* Canis *– but are different species. The red fox,* Vulpes vulpes, *is also in the dog family –* Canidae *– but is different enough to be placed in a different genus.*

population of greenfly on your runner beans may well be a clone, like
a patch of dandelions.

This kind of point is important ecologically. Asexual reproduction
is in general more rapid and more efficient than sexual reproduction,
and hence is commonly practised by creatures that benefit by rapid
colonization. But it does carry drawbacks, because animals and plants
that are genetically identical to one another are all equally vulnerable
to the same parasites, and so are easily wiped out by epidemic.

In addition, although asexual populations are capable of rapid
expansion, this does not necessarily mean that they will always reach
the largest population size. If it did, then all habitats would inevitably
be dominated by asexual reproducers, which is not the case. What,
then, determines final population size?

The size of populations: balance and minimum numbers

All creatures reproduce. Their ability to do so is one of the crucial
ways in which we recognize that they are alive; the ability to
reproduce is part of the *definition* of life. Reproduction is indeed vital
for the continuation of life, because any one individual is *bound* to die
or be killed at some stage, even if it does not have mortality built into
it, just as all buildings are bound to fall down sooner or later, and all
test-tubes are bound to be broken. Reproduction, providing a
constant stream of new individuals, ensures that at least some will
escape the particular disaster that overtakes their parent or parents.
Indeed the potential growth rate even of slow-breeding animals such
as elephants and humans, is phenomenal. Professor Ansley Coale of
Princeton University, New Jersey, has calculated that if the human
population had continued to grow at the rate at which it was
expanding in the 1960s, it would have reached 17 *trillion* – 17 million
million – within 700 years!

We know, though, that populations cannot continue unchecked. The
human population can never reach 17 trillion because there can never be
enough food to support that many people (or even enough oxygen for
them to breathe). Thus it was that at the end of the eighteenth century
the Reverend Thomas Malthus predicted imminent collapse for the
human race, precisely because food output could not keep pace with the
rapidly expanding populations brought about by the agricultural and
industrial revolutions. Charles Darwin took up Malthus's ideas, and
from them concluded that all creatures were bound to be in competition
with one another, simply because the world could not possibly support
all the individuals that were bound to be born.

We may argue with some of the details of Malthus's argument. We
may point out that the human species has not yet collapsed in the
way he predicted. Some biologists take issue with some of Darwin's
conclusions. But we cannot escape Darwin's central contention – that
all creatures are potentially able to produce more offspring than their
environment could possibly support. What stops them doing so?

Population: patterns and strategies

In some animal populations we can perceive a pattern that is commonly known as *boom and bust*. The population rises to enormous heights and then, for various reasons, collapses – often much more quickly than it rose.

With some animals – particularly small ones – this pattern of boom and bust is clear to see. Many insects rush through several generations in spring and summer, each one multiplying hundreds of times, so that in a few brief weeks even half a dozen flies could have given rise to millions. Small marsupials in the Australian bush commonly hide out in small, favoured pockets and then multiply, like flies, when the good times return. In the rest of the world, small rodents – rats and mice – follow the same pattern.

Usually, several factors combine to bring the boom to an end. The first is the rapid reproduction of predators, which cash in on the new food supply. Thus, when a wave of aphids descends upon a field of cereals, say, it finds a range of generalist predators already lying in wait for it: mainly spiders, ground beetles, and rove beetles, which feed on other things when there are no aphids to eat. As aphid numbers build up, so, too, do the numbers of various specialist predators that consume aphids exclusively, or nearly so. Chief of these are ladybird beetles, hover flies, lacewings, and parasitoid wasps. These last lay their eggs in the aphids' bodies – and the emergent larvae consume their hosts from the inside. A rapid flux of rodents is rapidly consumed by all manner of carnivores, from cats to lizards, falcons to wolves and coyotes. Any small animals that escape these depredations – as many surely will – are then sadly reduced by the onset of winter, or the return of drought. But a few survive: enough to start the cycle again when the summer or the rains return.

The animals in which we can easily see boom and bust in action are generally the 'r-strategists'; creatures that produce a great many offspring in a short time. Of course most of the offspring die; but a few pull through. The alternative reproductive strategy is known as 'K'; the animals produce only a few offspring, but each one has a high chance of survival. Typically, K-strategists produce large offspring (as opposed to tiny eggs or extremely immature babies) and often look after them for long periods. Primates, elephants, rhinos, whales, birds of prey and albatrosses are outstanding K-strategists. K-strategists don't necessarily produce large offspring, however. Bear and panda babies are notoriously small.

Because the populations of most K-strategists rise only slowly, it is easy to overlook the fact that in some of them too, populations may boom and bust, just as they so obviously do among r-strategists. But in the Ks, the cycles may take decades or even centuries to unfold. So they are very difficult to perceive – and in practice have to be inferred from historical or palaeontological records, or even just inferred on the basis of theory. However, boom and bust among K-strategists is now emerging as a highly significant ecological fact, with enormous implications for animal conservation and welfare.

Elephants could provide the prime example. In fact, at present, there are two contrasting views about the ways in which elephant populations are regulated. One of these is as follows.

If you observed an elephant herd in a state of nature for a few years then – in the absence of poachers – you probably would not see much change in numbers. You would very reasonably assume that however many elephants there were, was the 'right' number for that area, and that it would remain at that level.

But elephants have no outstanding natural enemies. Hyaenas will steal baby rhinos; but they have would have much more trouble with baby elephants, because elephants are sociable, and they gang together to ward off danger. Over time, then, however slowly, their numbers inexorably rise.

Elephants are also highly destructive. They are selective feeders, and they will break entire trees to reach some succulent branch. They also dig for water. If there are only a few, this cavalier spirit is of no consequence – and indeed is helpful, opening up new patches of forest, and providing watering places for hosts of other animals. But as numbers rise, they begin to damage more than can easily be replaced. Their environment starts to collapse. Once this happens, there is nothing for the elephants to do, but die.

So ecologists now conclude that elephant populations are not naturally stable at all. Probably for as long as there have been elephants in Africa and Asia, their numbers have risen inexorably over centuries until they were thick on the ground. But then there

▲ Elephants seem models of solidity. They are, however, extremely destructive (as this picture from a Zambian reserve illustrates) and the question arises, 'how has their habitat coped with their depredations over the past few millions of years?' Ecologists propose two kinds of answer. One is that elephant numbers have boomed and bust – not, like lemming populations, over periods of a few years, but over centuries or millenia. The other is that after elephants damaged one area, they simply moved on to the next – not to return until the first area had recovered. But each of these possible strategies is sustainable only when the animals have the entire continent as their living space. In a small reserve, elephant numbers must be regulated.

must have been periods of horrendous collapse – to be followed, over decades, by recovery of the environment; and then by an elephant resurgence.

But such a strategy 'works' only when the animals have all the time in the world, and when they have an entire continent as a stage. Nowadays elephants – at best – are confined to national parks. Some national parks in Africa are big, the size of small countries. Yet they are not big enough to sustain the boom and bust strategy of elephants. A period of elephant destructiveness, coinciding with drought, could permanently reduce the park to desert. Even in the best-run parks, then – or, rather, *especially* in the best-run parks, for in the worst-run ones they are simply killed by poachers – elephants have to be culled, to keep their numbers steady. Nobody likes doing this. There is a clear but inescapable conflict between the needs of conservation, and the demands of welfare.

Some biologists, however, do not agree that elephant populations naturally 'boom and bust'. They suggest instead that numbers probably remain fairly steady from decade to decade, and century to century, but that the animals move, effectively *en masse*, from region to region. Thus they would be behaving somewhat like traditional 'slash-'n'-burn', or 'swidden' farmers, who clear an area of forest, and then move on, but return later when the forest has recovered. In either case, the strategy works only when there is a vast area to move around in. 'Swidden' agriculture does not work if the farmers are confined to one spot; and elephant strategy – whether they are wanderers or boom and busters – cannot work if the animals are confined to parks.

It seems, then, that even within creatures that seem to be pillars of steadfastness, and even within ecosystems that seem (like tropical forests) to be models of stability, there are, even in undisturbed nature, huge fluctuations within the populations that make up those ecosystems. Yet traditional ecologists were wont to speak of the 'balance of nature'. In view of all the ups and downs, what weight should we now attach to that concept?

'The balance of nature'

Wherever you look in nature, you seem to see evidence of 'balance'. Carnivores eat herbivores, herbivores eat plants; all of them die, rot, and feed the plants again. If the plants increase, the herbivores will multiply too; but if they multiply too much they over-stretch the plants and their numbers fall again. Perfect balance indeed.

There are, though, some major caveats. First, a picture that looks steady when considered over a vast period of time, may none the less be enormously turbulent when looked on the smaller scale. Thus elephant populations in Africa may well have been roughly the same 1000 years ago as they were a million years ago. But there *may* have been enormous fluctuations in the mean time. The same applies when we contrast vast areas with smaller areas. Thus, one part of a biome (tropical forest, savannah, what you will) may look very like another. But if you look closely at any complex biome, you find that the fortunes of any one place within it may fluctuate: sometimes flooded, sometimes arid; sometimes teeming with life, sometimes bereft. In both cases, what maintains the illusion of steadiness is the enormous scale of time, and of space. Reduce the space (reduce the wild continent to a nature reserve) and the safeguards of scale are lost. The *natural* fluctuations of a decade or so can produce swings of fortune that will drive half the creatures (or more) to extinction. This is why wildlife management these days is a constant and uphill struggle: not simply to exclude poachers and pollutants, but to dampen the natural swings.

Neither is it true that the changes fluctuate around any particular point; that tropical forest is bound to remain forest, or savannah remain savannah, no matter how big the local turbulence. Dig down in almost any desert, and you will find evidence of forest beneath. Dig down in any forest, and you are likely to find evidence of desert (or of ocean bed!). Thus the fortunes even of entire biomes change with time. Neither is it true that the Earth itself is 'fixed' overall in any particular state. In the beginning, there was no oxygen. Now there is. The Earth contained no mechanism to prevent its influx. Life adjusted to it; but life and the Earth in general are very different now from the way they were before photosynthesis.

James Lovelock, who we have mentioned before, has argued that the whole Earth should be regarded as if it were an organism – 'Gaia' – which tends to regulate itself, just as any other creature regulates itself. In so far as this metaphor increases respect for the planet, it is grand. But if taken too literally it seems difficult to sustain: clearly, for example, the Earth had no mechanism to prevent the influx of oxygen. Taken too literally, also, it is dangerous. We may easily fool ourselves into thinking that 'life' can resist anything we care to throw at it: that sooner or later the 'balance' will be restored. But it ain't necessarily so. By our actions we can create a quite new Earth, in several different versions. But none is likely to improve on the one we inherited.

However, we should return to our principal theme, and consider another – often underestimated – form of population control.

Parasites

Running out of food (as elephants do at the peak of their booms) and being preyed upon are not the only checks on population. Parasites – which in this context means everything from viruses to tapeworms – also play an enormous part in regulating the size of animal populations. That this is so has been properly appreciated only in recent years. Before that, naturalists tended to nurse the romantic notion that wild animals were invariably healthy. The epidemic of phocine distemper in the North Sea, which in its first wave killed 17 000 common seals between April 1988 and January 1989 may have been exacerbated by pollution, reducing the seals' resistance. It is equally likely, however, that this outbreak was entirely natural, and that it would have happened even without human misdemeanour. Canine distemper recently all but wiped out the last of the wild black-footed ferrets in Wyoming. Outbreaks of all kinds of diseases – rabies, anthrax, bluetongue, Johne's disease, various tick-borne diseases, *Chlamydia* among koalas, scores of kinds of parasites in birds – have now been recorded throughout the world.

Infection in the wild has considerable conservational significance. Some animals are liable to catch human diseases – gorillas are prone to measles for example, and Jane Goodall's chimpanzees in Gombe (Tanzania) succumbed to polio, which could well have had human origins. It may become increasingly necessary to treat animals for disease in the wild. Vast herds of wild creatures can withstand heavy losses. But the diminished populations of modern times could all too easily be wiped out by epidemic.

Some animals seem to have built-in checks, which reduce their reproduction when populations rise too high. After a hard winter, when populations have been greatly reduced, many birds (such as starlings) lay more eggs than they do after a mild winter, when the spring populations are already high. We should not be romantic about this. The birds do not take an inventory and see how many starlings are 'needed' to maintain the great tapestry of life. The adjustment of breeding rate can be explained in purely self-centred terms. In general, it is in an animal's interests to produce as many offspring as possible; but if conditions are crowded, it will have a greater chance of successfully raising at least some offspring if it produces only a few in the first place.

In contrast, many animals breed well only if they are *stimulated* to do so by vast numbers of their fellows. Many colonial nesting birds,

▼ *Flamingoes, like many social animals, will not breed unless surrounded by other breeding animals of their own kind. Biologists at the Wildfowl and Wetlands Trust at Slimbridge, England, encourage flamingoes to breed by making them think the population is larger than it really is – by fitting their enclosure with mirrors.*

Disease in general; seals in particular

Between April 1988 and January 1989, 17 000 common seals were found dead around the coasts of the North Atlantic: a third of the European population. Pollution of the North Sea was at first suspected as the cause – and it may indeed have played a part in reducing the animals' resistance. But it now seems certain that the principal cause was a virus, distantly related to the measles virus of humans, and closely related to canine distemper in dogs; indeed, the infection is now called phocine distemper ('phocine' meaning 'relating to seals').

Whether or not pollution played a part in spreading phocine distemper, it now seems possible that the latest epidemic is not unique: some historical records suggest that there have been similar outbreaks every sixty years or so in the past. It is possible, indeed, that the North Atlantic common seals are merely illustrating a principle that ecologists have come fully to appreciate only in the past two decades: that infections and parasites play an enormous part in the lives of all wild creatures, animals and plants. They often determine, in large part, the size of wild populations; they often decide whether two species can live together, or whether one must give way to the other; and they have helped to shape the evolution of all living creatures – including, probably, the evolution of sexual reproduction.

Of course, many factors combine to determine the size of any wild population. North Atlantic seals in the absence of human beings are possibly limited by the availability of fish (though there would probably always be plenty of food for them); more probably by the availability of good breeding grounds (for they need hospitable beaches to produce their pups); and, to a small extent, by the depredations of their enemies, notably killer whales. In the modern world, of course, human beings probably do most to limit seal population size, by competing for food, expropriating beaches, and culling animals when we feel there are too many.

But the role of infection in limiting numbers, though very hard to determine, should never be underestimated. Though most infections and parasites do little damage most of the time, many are capable of reducing fitness (and hence reducing reproductive capacity) if the creature is weakened (perhaps by other infection) or if the parasite population grows. It is clear, for example, that the microbe known as *Chlamydia* is a constant blight in the lives of koalas in Australia. It is transmitted sexually and, among other ill-effects, it directly reduces fertility.

The role of infection in limiting populations is to some extent fine-tuned, because infection is liable to become worse as the number of hosts increases. Pathogens can spread more easily as the hosts become more close-packed, and large populations produce large numbers of young that are susceptible to the infection. Thus the constant presence of infection acts like the governor on an engine, cutting in when things get over-heated.

When conditions are right some infections are able to break out at intervals and reduce populations dramatically: more dramatically than any other natural factor is liable to do, short of a meteor strike or a tidal wave. Phocine distemper is probably a case in point. We know of many other epidemics among wild animals too: rinderpest in African game animals at the turn of this century, for example, brought in with European cattle.

The common seal epidemic illustrates, too, how infection may mediate in relations between species. Many pathogens and parasites live in comparative harmony with some host species – taking a small toll but doing very little harm. Yet those same pathogens may be devastating in another species. It now seems likely that the phocine distemper virus lives fairly harmlessly in the harp seal and the ringed seal which live in the Arctic, and that it will also tick along without too much trouble in populations of grey seal in the North Atlantic. But common seals are wiped out by it. The periodic epidemics among common seals may have occurred because, every now and again, harp seals choose to migrate south. The existence of phocine distemper means, though, that common seals would not find it easy to live in the company of harp seals, even if

such as flamingoes and gannets, are of this type. It seems that by breeding all together, such birds reduce the proportion of chicks that are lost to predators, such as herring gulls. After all, a gull can eat only a limited number of baby flamingoes in a given time.

The fact that animal populations often seem to stay reasonably steady, the fact that ecosystems tend to persist from year to year, even though there are booms and busts within them (as with flies in

▲ *Wild animals are not necessarily healthy. A third of all of Europe's common seals were wiped out by virus in 1988–89.*

there was no other competition between them. Similarly, red deer in Britain can be devastated by malignant catarrhal fever, which is carried by sheep (in whom it does no harm); and many species of hoofed animals in zoos are in danger from a variant of the same disease, carried by wildebeest.

The supreme significance of infection in the affairs of all creatures is illustrated by the now widely-accepted theories of Professor William Hamilton, of Oxford University. He points out (following John Maynard Smith, of Sussex University) that sexual reproduction is a highly unlikely process to have evolved: difficult, time consuming, dangerous (who'd be a male spider?), and inefficient, because sex requires two parents to produce each offspring, whereas asexual reproduction requires only one. The point though, says Professor Hamilton, is that sexual reproduction results in subtle variations from one generation to the next; variations that ensure that no single parasite can ever adapt fully to any one kind of host, and wipe it out. Sexual reproduction is extremely common: the norm among animals and plants, and practised by most fungi. That this should be so, in the face of all the trouble it causes, is a tribute to the power of infection.

an English summer), and the fact that most populations are naturally culled, just as surely as they multiply, has given rise to the concept of balance of nature. It is a reasonable concept, but as we have seen, it conceals a multitude of traumas. And, as is all too obvious, such balance as there is, is very precarious. All populations have their Achilles heel. All are vulnerable. And if the pressures begin to out-weigh the power to resist, the population goes extinct.

Extinction

It is the fate of all species to go extinct, just as it is the lot of all individuals to die. Most species that have lived in the past have gone by the board. Some people use this observation as an excuse to allow present-day species to disappear. Morally this is feeble – we could as soon condone premature death, on the grounds that everyone has to die sooner or later. But it is also bad biology. As the Amerian biologist Jared Diamond has pointed out, not all of the extinct species of the past were actually obliterated. Instead, many evolved into new species. Thus it is commonly believed that the extinct 'ape-man' known as *Australopithecus* evolved into human beings.

It is beyond question, however, that at least five times in the past 570 million years (that is, during the interval known as the 'Phanerozoic', meaning 'visible animals') living things have suffered *mass extinctions*. The best known of these occurred at the end of the Cretaceous period, around 65 million years ago, and saw the end of the dinosaurs. But the most extensive mass extinction was at the end of the Permian age, 230 million years ago, in which 50 per cent of all *families* of marine animals completely disappeared.

But human beings, through all their frenetic activities, are now perpetrating the biggest mass extinction of all. We have already cut down half the world's rain forests, which harbour at least 90 per cent of species on Earth. If we remove the other half – which at the present rate of destruction would happen within the early decades of the next century – then they would take most of life's variety with them. And although the mass extinctions of the past appear as sudden shifts in the fossil record, we know that each of them probably took millions of years to unfold. The present mass extinction, seen in biological terms, is taking place virtually instantaneously. For the first time, too, mass extinction is striking primarily at tropical forest, life's richest haven. The great Permian extinction took place mainly in the sea.

The extinction within the forests has obvious cause and effect: if their home is removed, then all the millions of forest dwellers are bound to perish. But extinction can creep up much more subtly than this. We may drive animals to oblivion even when we seem to be behaving reasonably responsibly, for reasons we should now discuss.

The vulnerability of living things

All creatures, as we have seen, are subject to set-backs: disease, predation, the onset of winter, drought, or bush-fire. Thus their populations may wax and wane. Provided the population is reasonably large, however, then it takes these fluctuations in its stride. But it can be shown purely by statistical analysis that if a populations of animals falls below about 50, and remains at such low levels, then that population is almost *bound* to be wiped out, sooner or later, by some natural set-back. In the modern world, many animals are already reduced to such levels. At the beginning of the 1980s, indeed, there were almost exactly 50 Puerto Rican parrots left in

▲ *Grizzly bears, like other big predators, need an enormous amount of space. Even mighty Yellowstone in North America, one of the biggest national parks in the world, may not be big enough to contain a viable population.*

Puerto Rico. As if to prove the statisticians' predictions, Hurricane Hugo came along and killed half of them. A few more disasters like that and they would be doomed.

However, the true picture is far worse than these crude figures suggest. Most animals, as we have seen, reproduce sexually. Most prefer to out-breed; that is, they need to mate with individuals that are somewhat genetically different from themselves (though of course, of the same species). If they do not, they begin to suffer from 'in-breeding depression' – which, in extreme form, results in obvious genetic disorder. A great deal of biological evidence suggests that unless populations contain at least 500 individuals, then some degree of in-breeding depression is almost bound to ensue (unless firm steps are taken to avoid this, in organized programmes of captive breeding).

With large animals, it takes a lot of space to maintain 500 individuals. Large carnivores in particular may command vast territories. After all, a sizeable herd of herbivores is needed to sustain just one carnivore. The last havens for wild animals on Earth are the national parks, but even the largest of these are hardly big enough for the largest animals. Even Yellowstone, mightiest of the North American parks, is hardly large enough for its grizzly bears; and even the apparently vast spaces of Africa's reserves are hardly large enough for wild dogs.

However, the extinction of a population may take years, even decades, to unfold. Thus we may gain the illusion that a particular animal is safe enough, only to find – when it is too late – that it is already fading. Cockatoos in Western Australia provide a fine example. Typically, they nest in holes in trees. Foresters have left plenty of trees in Western Australia, and there are still plenty of cockatoos around, so it seems there is nothing to worry about. But foresters clear away the old and rotten trees – which of course are the ones with the holes in. So there are very few places left for the birds to breed. Cockatoos are very long lived: perhaps fifty years or more. The ones born twenty or thirty years ago are still with us. But many of them probably have not bred for at least a decade, and when they finally begin to age, the population is liable to sudden collapse. The Asian elephant provides another example of the same phenomenon. Now (1990) there may be 30 000 left in the world, which seems a lot. But they are breeding only spasmodically.

Small nature reserves can also be highly illusory in the short term. We saw in Chapter 4 that islands cannot sustain a huge variety of creatures, because they are not large enough to sustain viable populations of more than a few kinds. But nature reserves are like islands. Pristine forests tend to be reduced to little copses, isolated among agricultural fields. Ecologists who return to such a copse the year after the main forest has been felled, may well be encouraged. They may well find a high proportion of the original 10 000 species remaining. But if they went back five years later, they would probably find that about half of the original types had gone; and within another five years, another half; and so on. Probably after a few decades, only a few hundred of the original types would remain. There seemed to be a great many in the early years, only because many animals take several years to die. This loss of creatures from a diminished habitat is known as *species relaxation* (see p.74).

Yet again, however, we have to admit that the true picture is even worse than we have so far described. Among the arable fields of Western Australia are thousands of *remnants* of the original bush, and of eucalyptus forest. In addition to species relaxation these remnants suffer *edge effects*; the encroachment, from all around, of outside influences. High among these is fertilizer, applied to the surrounding fields. Most wild plants do not appreciate high fertility. Weeds do, which is partly why they succeed in agricultural land. Relaxation is enhanced, then, by competition from vigorous invaders.

This kind of principle applies the world over. We can see the devastation of the rain forest, the marvellous trees reduced to scarred hillside and to parching grassland. But most of the landscape that looks so pleasing in Britain and New England, France and Australia, is largely whited sepulchre. Much of it cannot sustain anything like the number of species found in the pristine flora and fauna. The present mass extinction that we are perpetrating really is the biggest and most sudden of them all.

Adaptation

But do we not *define* life partly in terms of its ability to adapt? We are changing the world, but surely animals could learn to live in new circumstances?

Well, we have already seen some of the limitations of this argument; there simply is not enough space any more to sustain viable numbers of all the kinds of creatures that now exist. In addition, however, the changes we are making now are just too rapid. We are outstripping the ability of animals to adjust.

Adaptations can be of two main kinds. Some are acquired in the animal's life, and some evolve through natural selection, are imprinted in the genes, and inherited.

Short-term adaptations are often behavioural; and novel adaptive behaviour is often acquired during the course of an animal's life (that is, it is learned) or it may be imprinted in the genes.

Many birds can now be seen to be acquiring new forms of adaptive behaviour. Robins in the wild are timid woodland birds; but some have learned that there may be more to be gained by associating with human beings than by shunning them, and they happily sit on the gardener's spade. Carrion crows and wood pigeons, too, are impossible to approach in the country, yet wander through urban public parks. To some extent this may well be genetic; there may be sub-populations of especially tame urban robins. But in many birds such adaptation may be *purely* learned. Greylag geese migrate from Scandinavia to Southern Europe. In Danish parks they behave like tame birds – as Canada geese do in Britain. But once they get to Spain, they live as shy, wild birds. If the Spaniards catch them, they eat them.

▶ *Many animals are learning that it can be more advantageous to risk close contact with human beings, than to take flight. The robin is one such. Once, it was just a shy denizen of woodlands. Now, many survive in gardens by being 'tame'.*

Genetic adaptation takes longer. It *can* be rapid – but only if the animal is particularly 'lucky', and already has the necessary adaptation on tap. Peppered moths are normally grey and speckled, and are beautifully camouflaged when they rest on lichen-covered trees. But a few peppered moths are melanic – that is, they are almost black. We see the same phenomenon – a few melanic types – among leopards and jaguars. As the soot of industry began to kill the lichen on the trees in the North of England in the early twentieth century, so the peppery forms of the peppered moths gave way to melanics which were better camouflaged against the sooty tree-trunks. Now that the lichen is beginning to grow again, the pristine colour is becoming more common again.

Usually, however, animals have no such adaptation waiting in reserve. Usually it takes many generations to re-adapt to changed circumstance; and many lineages simply may not have the genetic or physiological equipment to make the necessary changes. We do not know precisely what killed the dinosaurs, but it seems to have been a climatic change, which apparently occurred over many thousands of years (though it may have been exacerbated by collision with some meteorite). Whatever the nature of the change, however, the dinosaurs clearly could not adapt.

The changes we are making today are, in general, far too rapid to allow significant genetically based adaptation.

Finally, in the light of all our comments so far, just consider the full implications of, for example, an oil spill. It is obliterative; it renders

137

huge areas of habitat inaccessible. Thus it reduces all populations with which it comes into contact, and would certainly render some of them locally extinct.

The effect is made worse because individual marine species tend to piece their habitats together from various components: plankton, intertidal zone, and so on. Damage to any one component will thus affect parts of the system elsewhere. Then again, all the creatures are inevitably locked into a food web. If one is affected, then all the others are affected as well.

In theory, perhaps, animals might adapt to the presence of oil. It is hard to envisage the kinds of adaptation that might enable them to tolerate it: perhaps a gut flora that would enable them to digest it. But they are most unlikely ever to evolve such adaptations; and even if this was biologically possible, a sudden oil slick obviously offers far too little time.

Just by living on this planet in such numbers, we make enormous inroads on our fellow creatures. Any degree of carelessness increases the problems that we pose a thousand fold.

On a more positive note, however, let us look again at how animals and plants relate to their fellows – of their own and of other species – in a state of nature.

Interactions: communities

Deep in the heart of ecology is the idea of 'community': collections of different species that apparently live harmoniously together, or at least in what seems to be balance, each adapted to some extent to the presence of the others. Wherever we look in the world, the idea that communities are real seems to stand up. True, the cast-lists differ from place to place – the animals of Pacific beaches tend to be somewhat different from those of Atlantic beaches; Amazon rain forest does not have the same trees as Malaysian rain forest. But if two places are similar in climate and topography then they will, by and large, attract and develop collections of creatures that have much the same general 'feel' as each other.

Indeed, the idea of the community should be taken seriously. Natural selection is the process through which living things become adapted to their environment, both physically and – to some extent – by their behaviour; and ecology is the study of the end result of that adaptation. A creature's environment always includes others of its own kind, and of other species – so each creature is bound to be shaped to a large extent by the creatures that surround it. Inevitably, then, within any one collection of creatures in any one place, there will develop powerful interrelationships. The different creatures will indeed be locked into a community.

However, we should acknowledge some caveats. The main one is that whereas some species are extremely specialized, others are equally versatile. The word 'specialization' has many meanings, of course. We can say that a bird is specialized for flying, or a tiger for killing. What is most relevant to the idea of the community, however,

► *Life is a balance between co-operation and competition. Grassland antelope benefit from living in herds, because while some are feeding, others will be on the look-out. Many predators, such as lions, co-operate in hunting. But when the predator strikes, it may be every antelope for himself; and it is often the weakest that is singled out, and caught.*

◀ *Animals can adapt to change if given time. But many of the changes wrought these days by human beings are too dramatic, and too rapid, to allow such adaptations. This bird has been killed by an oil spill.*

is one extreme form of specialization, in which one particular creature becomes totally reliant on the existence of some other particular creature.

Such reliance takes various forms. Sometimes one creature simply lives on, with, or inside another without doing it any particular harm, or any particular good, and is then said to be a *commensal*. Many of the bacteria that live in the guts of animals are of this type; such as the *Escherichia coli* that live in the guts of human beings. Many different kinds of plant live on the trunks or branches of trees. These plants that live on other plants are called *epiphytes*; and they too are examples of commensals. Epiphytes come from many different classes of plant, and include liverworts, mosses, ferns, and flowering plants including many orchids and bromeliads (which are relatives of the pineapple, and include the Spanish moss of the southern United States).

Predators may come to rely upon only one (or only a few) other species. Thus koalas are almost totally committed to the existence of eucalyptus trees, and aardvarks depend upon termites. Parasites also tend to be highly specialized to their hosts. Thus some parasitoid wasps lay their eggs only within one (or a very few) species of aphids.

Often, however – remarkably often, indeed – different creatures become locked into a very close relationship from which *both* participants benefit. Such a relationship is said to be *symbiotic*. Thus, various algae co-operate with various fungi to form lichens. Bees have a symbiotic relationship with flowers which they pollinate, obtaining nutrient-rich nectar in return. In temperate countries, the bees tend to be versatile; they will feed on all the flowers of the meadow, or on heather or clover, depending on what is available. But in tropical forests many species of plant are pollinated only by particular species of insect. Thus in Amazonia there are at least 80 species of fig, each pollinated by its own species of wasp. One eighth of all flowering plants are orchids – 30 000 species out of 250 000 – and many of them,

in tropical forests, are pollinated by their own particular insects.

There are many other symbiotic relationships that are not so easy to observe, but which are vital to the smooth running of the whole biosystem. Herbivorous animals could not subsist on coarse vegetation such as leaves without the help of symbiotic bacteria in their guts (see p.108). Forest trees of all kinds, and many other plants as well, fare badly unless their roots are first invaded with symbiotic fungi to form associations known as mycorrhiza. The fungi enhance the absorptive powers of the roots, and in turn receive organic molecules from the tree. The fertility of the world's soils would be sadly compromised were it not for the existence of nitrogen-fixing bacteria, which capture nitrogen gas from the air and turn it into soluble nitrogenous materials which plants can absorb (see p.104).

Within any one assemblage of creatures in any one place, then, you will find a great many close relationships: specialist predators feeding on specialist prey, commensals, specialist parasites, and dozens and dozens of symbiotic relationships. There will be many overlaps, too. Most animals and plants are preyed upon by several different specialist parasites (it is reckoned that each species of vertebrate has at least one specialist roundworm – nematode – to feed upon it); and some predators and parasites, though still qualifying as specialists, may well prey upon more than one host.

Put all the relationships and cross-links together, and we find that each assemblage of creatures in any one place may indeed form a tight community. We find, furthermore, that if any one is removed,

▲ *Two striking and common examples of symbiosis. Lichens (above left) are co-operatives between fungi and algae. The algae photosynthesize, and provide organic materials; the fungi provide a safe haven, moisture, and help absorption of inorganic nutrients. Most flowering plants and conifers benefit from fungi known as mycorrhizas which surround or invade their roots (above right). The hyphae of these fungi grow rapidly, spread into all the nooks and crannies, and increase the ability of the roots to take up nutrients; particularly phosphorus. In return, the fungi take organic materials produced by their host. Some plants need their mycorrhizas. Some prefer to have them, but can survive without.*

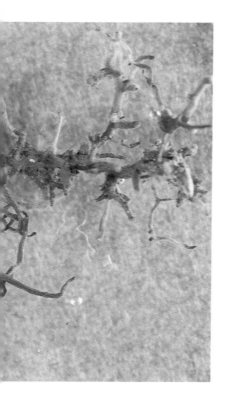

then many others that are dependent upon it may disappear as well; though others, that were formerly in competition with the one that has gone missing, or were preyed upon by it, may flourish – and they in turn will affect others. Thus, when any one species is removed from a community there may be a knock-on effect, like the ripples from a stone dropped in a pool. It is for this reason that complex ecosystems - such as tropical rain forests – are so fragile. There, the interdependencies are legion, and hundreds of species may disappear if only a few are removed.

Despite all these comments, however, we should not be carried away by the idea of community. Many creatures are extremely versatile. Common oaks, *Quercus robur*, will grow in a variety of places, and turn into spreading behemoths, or reach up straight for the sky, or remain as twisted shrubs, depending on circumstance. Tigers in India may hunt barasingha (which are large deer) in the swamps; black-buck (small antelope) in more open country; or nilgai (large antelope) and gaur (the world's largest species of wild cattle) in the hills.

Communities are thus usually more flexible than has sometimes been made out. If any one community is altered (for example if trees are selectively removed from a forest) then the remaining creatures will, in general, form themselves into a new community that is no less stable, and contains as great a biomass, as before. Or if some new

▶ *Many species are extremely versatile, and can survive in a wide range of environments. The oak is one such – though in harsh conditions, it may be twisted into bizarre forms.*

141

creature is introduced into an apparently stable community, it may remain to flourish, and may (with luck) become integrated to the point where future generations forget it is a newcomer. Thus rabbits have become an accepted part of the English fauna, although they were brought to Britain only about 2000 years ago (by the Romans); and dingoes are now considered native Australian, though they were probably introduced by a late wave of Aboriginal people, around 3000 years ago.

The flexibility of communities has good aspects and bad. What is good is that nature is indeed resilient. Cyclones batter at tropical forests or at coral reefs; but the reefs come surging back. Human beings do enormous damage to natural communities of all kinds, but generally, at least some collection of creatures usually manages to flourish in the changed circumstances.

The down side, however, is that we all too easily fool ourselves. We take vast sections out of a forest, see that it regenerates, and then convince ourselves that we have done no harm. Europeans have introduced scores of animals and plants into Australia and have perceived that they have often done well; yet introduced foxes, and domestic cats, have played havoc with the small native marsupials.

In short, by our interference we continue to create new communities the whole time, which may *look* very acceptable and which may indeed be green and pleasant lands. Yet those novel communities may lack many or even most of the creatures that were present originally, and contain others that may have originated in different continents. Such novel communities may be stable enough ecologically, and they may meet our own desire for a pleasing landscape. But such depleted and altered communities may in reality be relics: all that is left after mass extinction has already taken place.

Even in a state of nature, however, no community lasts *forever*. Some may persist for millions of years; but the fossil record shows that all good things come to an end. Most landscapes too are less stable than is immediately apparent: coastlines spread and recede, mountains erode, rivers find new ways through, islands sink, and so on. Any such physical change will prompt a change in wildlife, and many ecosystems, both natural and man-made, are in the process of changing to something else. We should look, then, finally, at the idea of change.

▲ *'Exotics' – animals introduced by human beings into new environments – often devastate the native creatures of their new home. Dingoes were probably brought from Asia into Australia by Aboriginal people, about 3000 years ago; and they drove the native Thylacine 'wolf' to extinction on the mainland. But foxes and cats, brought in by Europeans in recent centuries, have been far more destructive.*

Change and succession

Ecologists noted some decades ago that communities of living things do tend to change, as time passes, into communities with a different makeup. They noted, too, that the changes in any one kind of environment tend to follow a similar pattern; indeed, that within any one country or continent, particular species tend to enter the community, and perhaps displace others that were there before, in a more or less predictable sequence. They called this observable parade of species, *succession*. They noted, however, that eventually each

succession in each kind of habitat finally came to a halt, that a point was reached at which the vegetation simply went on replacing and replenishing itself, with no change of species, and that the fauna also, inevitably, reached a point at which there were no more newcomers, and no more extinctions. This final stage, in any one place, was called the *climax* vegetation.

The ideas of succession and climax are valid, and do indeed help us to understand the environment. Thus we can predict that heathland with birch trees will, in the fullness of time, probably give way to oak; and that oak forest will emerge as the climax vegetation in temperate lands that are not too dry (as it did, in the past, over most of Britain). Succession is not a simple idea, however, and – as with the idea of community – we should add some caveats.

First the term 'succession' is used to cover slightly different phenomena. Thus, after each of the great Ice Ages, the ice at first moved north, and the tree line followed shortly behind it. The first trees on the path north tended to be pines, alders, aspens and birches, with oaks following after. What we have here, however, is first, a gradual amelioriation of climate, which allows the cold-tolerant trees to invade first, and second, the fact that birches grow more quickly than oaks, and so become established more quickly.

The vegetation that prevails in a habitat at any one time tends to change the conditions in ways that favour different species. Thus vegetation (and the accompanying animals) tend to undergo a 'succession' until a 'climax' is reached, and no further succession takes place. The natural climax for most of Britain is oak forest.

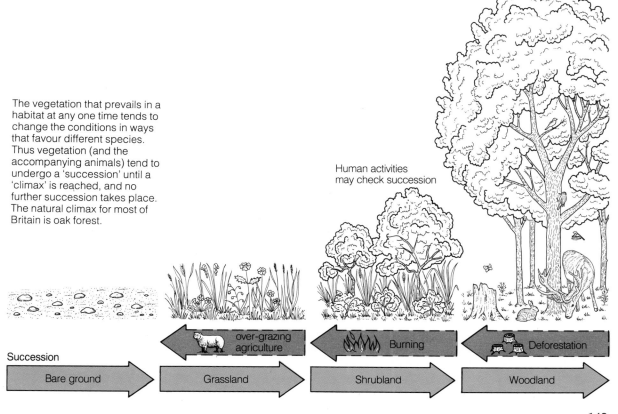

Human activities may check succession

| Succession | over-grazing agriculture | Burning | Deforestation |

| Bare ground | Grassland | Shrubland | Woodland |

There is another kind of 'succession', however, in which the climate stays the same, but the plants themselves alter the conditions, in such a way that each community creates conditions that favour a different set of plants. Thus in some salt-marshes, salt-loving grass such as *Spartina* may raise the soil level as it grows, which allows less obviously maritime grasses such as *Festuca* to invade. Unless the *Festuca* is grazed, then shrubs will start to invade the rapidly establishing soil, and continuing rain will reduce the salinity, and eventually forest trees again take over. Such sequences occur in clearings in tropical forests, too – clearings caused naturally by the collapse of a single tree, or by a storm. Some forest plants cannot germinate unless they are exposed to the sun – which happens when a clearing opens. But as soon as they begin to grow, they blot out the sun, and create conditions that favour plants that germinate only in the shade.

The first kind of succession, in which climate or landscape gradually changes, can operate in two directions; so if a new ice age began to descend, then the oak forests would gradually retreat before it, giving way first to woodland that was more cold-tolerant, and then to tundra, before disappearing beneath the ice again. But the second kind of succession does not happen in reverse, unless some physical change is imposed from outside. Oak forest does not naturally create conditions in which oak trees cannot grow. That is one reason why it is indeed climactic.

As with the idea of community, however, we should not be carried away by the idea of succession. It is true that in natural ecosystems in, say, northern Europe, particular groups of species do tend to replace other particular groups in a highly predictable way. But one reason for such predictability is simply that there are only a finite number of species available in any one locality. We should not conclude that any one particular sequence is inevitable, down to the last detail. If the environment were given more options, then it could often do things in a different way.

Thus at Stanford University, in California, James Drake set out to see if communities always did change by succession until they reached a particular climax, as ecologists have often implied. He built an artificial community, adding various species of aquatic creatures to aquaria. Always he added the same species – but he introduced them in different orders. If the classical ideas hold, then it should not have mattered too much what order he put them in. Eventually, the different creatures should have adjusted one to another, until they all reached the same balanced 'end point' – the same climax. In fact, this was not the case. The stable end point that eventually developed in each case depended very much on the sequence of introduction. In a pond – as opposed to a forest recovering after an ice age – the sequence of introduction of species may well be highly variable; and Dr Drake's studies suggest that what any one pond finishes up containing is very much a matter of time and chance.

In these seven chapters, we have described the principal habitats on Earth, and discussed the main ideas that ecologists have arrived at in their attempts to explain their workings. In the last chapter we will look at the effect, on all those natural systems, of human beings.

8 *The impact of people*

As we saw in Chapter 6, there are good ecological reasons why big predatory animals are bound to be rare. Yet humans are big – bigger than at least 99.9 per cent of all other animals. We are also predators, albeit omnivorous ones. Already, though, there are five billion of us. Among land vertebrates, only domestic chickens (which exist because of our efforts) and rats (which tag along behind us) are anything like as numerous. We have apparently defied one of the most fundamental of ecological laws. How?

The answer, of course, lies in agriculture. Wild soil – by agricultural standards – is generally not very fertile. Where there is adequate warmth, light, and water, there is likely to be a lack of nitrogen or phosphorus or perhaps some other nutrient – relative, that is, to what plants could, in theory, make use of. But in agriculture, we increase soil fertility and enhance the water supply, until plant growth is limited *only* by light and temperature; and in glasshouse horticulture, we may remove even those limitations. In addition, we grow only those plants we choose to eat, and breed varieties that flourish in highly fertilized soil. Then we breed animals that grow rapidly when they are overfed – by the standards of the wild, that is. In these ways we can and do increase the food output of pristine wilderness a hundred or even a thousand times.

We continue to improve agriculture. We continue to raise the fertility of more and more land, to reduce the toll of pests and diseases, and to develop crops and livestock that are more and more responsive. Accordingly, the human population continues to grow. How long can the expansion continue, of food output and of people? And will it inevitably end in disaster, or can we stop the increase gently, and finally establish stability?

Human population

When everyone on Earth lived by hunting animals and gathering wild plants, the maximum human population was probably around eight million – somewhat less than that of modern London. It probably stayed at around that level for tens of thousands of years until around 10 000 years ago, when farming began in earnest in the

Middle East. Early farmers undoubtedly had a hard time; successful hunting and gathering is a more agreeable way of life. But the name of the game is not pleasure, but survival, and farming can support more people. By the time of Christ, 2000 years ago – when Eurasia was already urban as well as agricultural, and farming was also underway in the Americas, Africa, and Australasia (in New Guinea) – the world population was probably around 300 million. Thus, there had been roughly a 40-fold increase in 8000 years.

Between AD 1 (the time of Christ) and AD 1750 the world population roughly tripled, to around 800 million. But it was then that agriculture began to become 'scientific', when western methods of agriculture were becoming established in other continents, and when agriculture began to be abetted by modern industry. After 1750, then, the growth of the human population got truly into its stride. By 1850 the world population was around 1.3 billion; by 1900, 1.7 billion; by 1950, 2.5 billion; and the 5 billion mark was passed in the late 1980s.

The figures themselves show that the rate of growth has accelerated; but it takes a little analysis to see just how rapid that acceleration has been. Such analysis reveals that up until about 10 000 years ago, it took about 35 000 years for the human population to double in size. Between AD 1 and 1750, the doubling time for the human population was around 1200 years. At the present rate of increase, the human population could double in about 35 years. This means that by AD 2025 – not so very far away – there could be 10 billion people on Earth; 20 billion by AD 2060 – well within the lifetimes of many of today's schoolchildren; 40 billion by the end of the twenty-first century – when a few of today's babies should still be around; 80 billion by AD 2120 – when the grandchildren of some of the Earth's present inhabitants should still be living. In 350 years, at the present rate of increase, the world population would reach 4 *trillion*: 4 million million. Within 700 years (not so very long, even to a historian, and the twinkling of an eye in biological time) the human population would, at present rates of expansion, reach 17 trillion.

Some people welcome this increase. They feel that human beings are the pinnacle of creation, and that the more there are of us, the better. Others feel, however, that human beings should more humbly regard themselves simply as a *part* of creation, a species with many special qualities, to be sure, but one that should see itself as one species among many. Many feel, too, that if life becomes a simple struggle for survival, then it is not worth living, and that a simple desire for human increase is far from humanitarian. But it is clear, too, that continued expansion is not an option. Perhaps the world could support twice the present number of people, at least for a time. Perhaps it could just support four times as many (which would be reached at present rates by the mid-twenty-first century). But even the most optimistic technologists would doubt if it could support, say, 40 billion – the figure that could be reached in 100 years; and certainly not 80 billion – projected for our grandchildren's time.

Beyond question, something is going to happen within the next 100 years (within many present lifetimes) to slow the present growth.

▲ *People, people everywhere. This is Kyoto, Japan.*

When? How big will the population be before the reversal begins? And, will the reversal be voluntary and painless, or will our species end in disaster – epidemic, famine, global war? Whatever outcome we imagine for ourselves, we should not fall into the trap of concluding that we are immune from disaster. Think of the elephants, those invulnerable creatures whose populations crashed many times in the past. Think of our own history; the numerous peoples who have been wiped out completely in historical times, for example in Tasmania and the Americas, and the tremendous reversal in Europe in the fourteenth century, when millions died from plague. The fact that some events are too horrible to contemplate does not, unfortunately, mean that they cannot occur.

How many?

If human population does finally crash because of disaster, then projections into the future are nonsensical. Once events get out of hand, then predictions are impossible. Some scientists, notably Carl Sagan of Cornell University, New York, have projected that the change in climate (nuclear winter) that would follow a nuclear war could eliminate the entire human species, and although this is clearly an extreme view, it does at least show what is theoretically possible.

However, the human population could come down without disaster. Though human beings are highly variable and capricious, and we cannot predict their behaviour precisely, there is evidence that when people become reasonably well off and have plenty to occupy their time, and when they feel confident of the future, and feel reasonably certain that all of their children will survive into adulthood, then they are happy to have fewer children. If people average fewer than 2.3 children per female, then the population does eventually begin to go down (because some of those 2.3 will fail to have children of their own). If two children or fewer becomes the norm, then the decline becomes obvious. West Germany in recent years has shown what can happen. There, numbers started to go down, simply because people chose to have small families.

Demographers, for example at the Office of Population Research at Princeton University, New Jersey, suggest that if all the world followed West Germany's erstwhile lead, then the rate of rise in world population would begin to diminish almost immediately, and the population could stabilize at about 8 billion by about AD 2030. Demographers do, however, take into account the principle of *momentum*. That is, populations continue to grow (or to remain steady) even after the birthrate goes down, because the children who are already born begin to grow up and have children of their own. Because of momentum, then, the population would remain at about 8 billion perhaps for around 500 years, even if the average birthrate stayed at 2.1 per female or less. Then, though, it would begin to diminish of its own accord. Once the diminution was in train, future generations could simply decide how many people they would like to see on Earth. Paul Ehrlich, ecologist at Stanford University,

149

California, suggests that a world population of around one to two billion would be comfortable; this could in principle be achieved entirely by voluntary means within about 1000 to 1500 years. Michael Soule, ecologist at the University of Michigan, suggests that 100 million would be a comfortable world population; the number at the time of Christ and, as he points out, ' a time of towering genius'. Future generations can decide this, however. Note, though, that both Ehrlich and Soule are humanitarians. They are not suggesting that human beings should commit hara-kiri. They are merely suggesting that if we exercise some biological restraint, then future generations can hope to live in security and happiness. Without such restraint, the future is very uncertain indeed.

There can be little doubt, though, that the present population is not easily sustained. Certainly, it is possible to demonstrate with back-of-the-envelope calculations that present-day agriculture could in theory support 20 billion people or more (provided we all modified our present diet somewhat). In practice, though, agriculture can never work at full efficiency worldwide, for many political and social reasons, as well as technical ones. Neither do such projections take into account firstly the near certainty that the world's climate will be changed by the greenhouse effect, and that this will disrupt agricultural output in the short term and lose a great deal of land as sea-levels rise, and secondly the destruction and degradation of the environment. There are signs of hope of course: some rivers are being cleaned up; Europe (including Britain) has decided that it produces more food than it needs and is taking some land out of production

▲ *Enormous human populations can be supported only by intensive agriculture. In much of Asia, every square inch is spoken for. These are terraced rice fields, in China.*

150

(the policy of 'set-aside') which means that more land could be used for conservation; and national parks are still being created. Yet there is no doubt that the general degradation of the world environment is proceeding as rapidly as ever, and that the mounting human population is not making things easier. We may legitimately doubt, then (as Professors Ehrlich and Soule do) whether the world could sustain five billion people indefinitely – let alone the eight billion that seem bound to exist within a few decades. Michael Soule, indeed, talks of the 'demographic winter'; those centuries to come in which the human population will be unsustainably high.

With these thoughts in mind, many ecologists now feel that conservation must be seen as a holding operation. The task is to sustain as many of our fellow creatures as possible (as well as ourselves!) through the next 500 to 1000 years. After that, human numbers might have diminished enough to take some pressure off the environment.

What, in general, are the kinds of things that must be done?

Well, the greatest threat to other creatures (and, in the long term, to ourselves) is the simple loss of habitat. The main, though by no means the only cause of such loss, is the spread of agriculture. First of all, then, what can be done to ameliorate this?

Reducing the impact of agriculture

Many people realize that modern agriculture has become extremely damaging to other species, and suggest that the proper solution is to devise agriculture that is, in effect, 'less modern'. Others seize upon the fact that we use as much land as we do for food production, partly because we feed so much of what we grow to livestock. The way out, they suggest, is for us all to be vegetarian. There is sense in such ideas, but they should not be taken too literally. The changes in agriculture, and in diet, need to be more broadly based than these philosophies suggest.

Thus, although it is true that old-fashioned farms were generally more hospitable to wildlife than modern ones, it is not an option and neither is it desirable simply to reinstate old-fashioned methods. For one thing, old-fashioned farms were *not* entirely unpolluting. Nitrogen from manure applied to crops could 'run off' and pollute groundwater just as easily as modern fertilizers sometimes do. What counts most of all, is the amount that is applied, and the timing of application.

More serious, though, is that old-fashioned farms were far less productive than today's. Even forty years ago, wheat yields in Britain were only a third per unit area of what they are today, and at the start of this century, only one-sixth. So if we still farmed the way our great grandfathers did, we would need to use several times as much land to produce the same amount.

But agricultural science today, particularly as carried out by Britain's Agricultural and Food Research Council (AFRC), is intended specifically to provide agriculture that is both productive *and*

◄ *Pigs forage for acorns in England's New Forest. Traditional farming alone cannot feed the present world population, but some of its practices are both productive and humane, and deserve to be reconsidered.*

'environmentally friendly': for example, to devise methods of fertilizing crops for high yields, *without* run-off, and to find ways of controlling pests without using vast quantities of damaging pesticide. Only by such science – which is extremely modern, but none the less benign – can we hope to feed ourselves without continuing to damage our fellow creatures.

Neither does vegetarianism offer quite the panacea that it seems. It is true that in western countries we feed more grain and pulses (beans and peas) to livestock than is ecologically desirable (about half of all the cereal grown in the United States and in Britain, and more than two-thirds of the pulse). It is true, too, that human beings could derive all the protein and energy they need from eating cereal and pulse – so our meat animals are to a large extent *competing* with us for food.

However, no crop is edible in its entirety. Cereals produce straw and 'tail corn', of inferior quality. Those 'wastes' can be fed to animals. And not all land is suitable for grain and pulses. Britain has vast expanses of hills, where the best crops are grass and heather; no food for humans, but excellent for sheep, cattle, deer, and goats. In short, we could feed more people if we *cut down* on the amount of meat we eat. But we would make better use of our land and our crops if we allowed ourselves *some* meat than if we cut it out all together. Besides, meat is a source of nutrients other than protein and energy, including minerals such as zinc, and essential fats. It would be irresponsible to recommend strict vegetarianism to everybody in the world.

The ecological task for agriculturalists is three-fold. There is a continuing need to devise systems of agriculture that combine food production with conservation, including, for example, some use of traditional meadows where cattle graze profitably on hugely varied pasture, or pigs grow fat among forest trees. But there is also a need to grow crops ever more intensively (though without polluting the surroundings!) so that we can feed ourselves on less space and release more land for wildlife. The third task is to create an overall design, a mosaic, which strikes the best possible balance between extreme intensiveness, wilderness, and combined methods in between.

'Wilderness' though, cannot unfortunately be quite as wild as in the past. The world's remaining animals and plants must for the time being be crammed into spaces that are far smaller than the pristine continents. These spaces – 'reserves', 'national parks' – need to be managed if they are not to decay.

National parks

National parks should in general be as large as the countries that contain them can possibly afford. No ecologist doubts that. Size, though, is not all that counts. Most animals, as we have seen, compound their habitats from several different components of the environment. Big grazing animals need to move from pasture to

pasture as the seasons unfold. Ocean-going fish may breed in coral reefs or mangroves. It is important, then, that national parks should each contain as many as possible of the necessary components.

When the park is established, too, it must be knowledgeably managed. Forest trees should not be cleaned up too much: holes must be left for nesting birds, such as cockatoos. Some animals (including elephants) may need to be culled at intervals, and in a strictly controlled fashion, to stop them over-populating their environment. Encroaching weeds may have to be actively excluded. Sometimes it may be desirable to treat wild animals against infection. Today's populations probably cannot withstand epidemic the way they did in the past, simply because there are fewer individuals.

We are forced to acknowledge that, at least for the next few centuries, the space available strictly for wildlife will be limited. Ecologists worldwide are asking, 'is it better to put all the available land in one place (and call it a 'national park'); or to have many, smaller areas? If there are many small areas, should these be linked by 'corridors', or left separate?'

Each pattern of land has its own advantages and disadvantages. If there is one big area, then the ratio of edge to centre is reduced, which means there is proportionately more untrammelled habitat. But one big area is unlikely to cover the complete range of habitats in any one country, so there should probably be several areas in different regions. Links between different areas could be good, because they could help small populations of animals to merge together to form one large one; and one large population may survive better than two small ones. On the other hand, predators and diseases could also travel along linking corridors, and small areas, though intrinsically vulnerable, may none the less find relative safety in isolation. However, when the greenhouse effect begins to affect the climate, animals may need to migrate out of present national parks (which in some cases might simply become too arid, for example), in which case they will need corridors to migrate along.

Such dilemmas as these can be resolved only by good scientific research. Not enough is being done. These days, few governments spend significant sums on research that does not bring immediate commercial rewards.

No matter how well national parks are organized, however, for the next few centuries they will find it difficult to accommodate all the animals that need to live in them. They will have to supplemented, then, by populations of animals in captivity.

Captive breeding and reintroduction

Until about twenty years ago, or even ten years ago, few zoos believed that they could make a serious contribution to the survival of animals by breeding them. Certainly, the world's best zoos believed they had a part to play in conservation – but only for carrying out research and by bringing the plight of animals to people's attention. They did not seriously think they could keep animals in sufficient numbers to aid in their survival.

Now, however, ecologists realize that many wild populations are already below the 'magic' figure of 500 (see p.135), at which they can be considered reasonably safe. They all agree that habitat protection is the ideal solution, but in many cases, for the time being, this simply is not an option. The scimitar horned oryx was recently wiped out of one of its last strongholds in the Sahel by the civil war in Chad. One of the last populations of Puerto Rican parrots was recently halved by Hurricane Hugo. No-one can legislate against that. It is possible to legislate against poaching; but even so, black rhinoceros continues to be slaughtered through most of its remaining range – except for the animals on private ranches or in small, specialized reserves.

So, many of the world's leading zoos, especially in Britain, the

United States, Europe, and Australasia, are now co-operating in what the Americans call 'Species Survival Plans'. No one zoo can contain viable populations of animals such as tigers or black rhinos, but by judiciously exchanging animals (without exchange of money!) and by organizing the breeding according to sound genetic principles, they can treat their joint collections as one giant breeding population.

Captive breeding cannot provide the solution for all the world's endangered animals. If we include all the specialist beetles of the rain forests, after all, there are several millions of them; and the world's zoos are not that big. Zoos can, however, save several hundreds or even several thousands of creatures from immediate extinction. That is a small proportion of the whole, but it is a lot better than nothing, and those hundreds could, in principle, include virtually all of the larger vertebrates, from Siberian tigers to Grevy's zebra, whose immediate future in the wild looks grim.

Captive breeding does not represent the end of the line for these creatures. Eventually – when the human population has begun to settle down again – it should be possible to return many of them to the wild. In fact, reintroduction is already taking place, with examples ranging from Arabian oryx (into Oman – see p.158) to the red wolf (into Carolina). Worldwide about 100 reintroduction programmes are already in train.

Zoos depend upon public support; and reintroductions are bound to fail unless local people make the returning animals welcome. Thus, the Arabian oryx programme (the most successful so far, worldwide) relies upon support from the sultan of Oman, and the Harasis people, who act as wardens.

In general, though, public support for conservation is forthcoming only if people perceive that they receive some benefit from wildlife. This is why so many ecologists now seek to reconcile conservation with some kind of exploitation, to provide an immediate pay-off.

Conservation by exploitation

Exploitation of wildlife can take many forms. The most widespread, and obvious, is to use wild animals and plants as a means of attracting tourists. At the other extreme, it has often been pointed out that many of Africa's game animals fare much better than domestic cattle do in the harsh climate, and could be culled for meat. In between lie various schemes to allow limited hunting or harvesting of wild creatures, often combined with some artificial enhancement of their numbers. All approaches have their advantages and problems.

For example, tourism is undoubtedly lucrative. Kenya derives more than a third of its income from it, and most of the tourists come to look at the animals. But tourists put severe pressure both upon the animals and on the landscape. Vehicles tracking cross-country destroy native plants. Land-rovers crowd around cheetah kills, and they in turn attract hyaenas and jackals to the scene, which drive the cheetahs away. Richard Leakey, head of Kenya's Wildlife Service, is now seeking to raise about 150 million US dollars to improve Kenya's

▲ *There are still tens of thousands of Asian elephants left in the wild. But they are not breeding well, and in the next century the species may need support by captive breeding. So far, however, very few elephants have bred in zoos. Many zoos are happy to keep females, like this amiable animal from London Zoo. But few are equipped to keep bulls safely!*

wildlife parks – better but fewer vehicles; better roads; more guards and wardens – to ensure that the tourists get their money's worth, while the animals are left in reasonable peace.

There are many other exciting possibilities. In Queensland, Dr Graham Harrington, director of the Tropical Forest Research Centre, would like to create walkways through the forest canopy, for tourists to admire the animals and plants – which would surely be a delight. Similar schemes in Brazil and South-East Asia could yet do much to save their forests too. Tourism could be worth a lot more – and be a lot more permanent – than logging and farming.

Killing wild animals for meat is problematical as a conservation strategy. If the animals are culled from the wild, then there are problems both of hygiene and of logistics – for how can the carcasses be brought economically from the bush to the cities, where they are needed most? But if the animals are kept in semi-confinement, to make them easier to harvest, then they begin to lose some of their advantages. For example, they cannot range so freely in the cool of the night, and so cannot graze so efficiently. In general, culling of wild (or semi-wild) animals would seem to have only very limited possibilities, although in South Africa kudu and springbok are farmed on extensive ranges, and are commercially very profitable.

Limited hunting for trophies also has its place, but also raises difficulties. Thus in New Guinea, local people encourage wild butterflies by planting the trees that their larvae feed upon. They then sell the butterflies to tourists – and this gives them an interest in preserving the forest, albeit with a few extra trees in it. In Kenya, however, Richard Leakey is very much against the harvesting of ivory, even though surplus elephants have to be culled. The selling of ivory encourages the trade, he says. Once trade is established, the poachers redouble their efforts. The only way to stop poaching, he believes, is to persuade people that it is bad to own ivory – just as they already believe that it is bad to own the fur of endangered cats. He has evidence that this is the case, and would prefer to burn ivory (as the Kenyan government did in 1989) than to sell it, even though it fetches a high price.

There can hardly be a more urgent task in all of conservation – which means there can hardly be a more urgent task of any kind – than to reconcile human aspiration with the survival of wild animals and plants. Ecologists, experts in tourism, politicians, educators and financiers must combine to address and solve the problems.

In the future, though, wild animals and plants cannot be seen simply as an adornment for our leisure hours. If we truly want them to survive in large numbers, we must seek ways of accommodating them in all of our activities. We have mentioned the possible scope in farming. But we should welcome wild animals into our cities, too.

The Arabian oryx: model for the future?

The world now contains about 4000 species of mammals, 8000 birds, over 5000 reptiles and 2000 amphibians. Of all these, about 2000 are under some threat of imminent extinction. William Conway, Director of the Bronx Zoo at New York, has calculated that if all the world's zoos co-operated, then they could between them sustain about 800 species of land vertebrate by breeding them in captivity; some way short of what is desired, but still a significant contribution. Furthermore, the aim would not be to maintain these species as zoo animals in perpetuity, but to return them to the wild as soon as their native habitats can be made safe for them again.

To some people, plans to save animals by captive breeding followed by reintroduction still seem fanciful. So far, indeed, there have not been many successful examples. But the case of the Arabian oryx shows what can be achieved.

The Arabian oryx is a magnificent animal which once lived throughout the Arabian peninsula – an area the size of India – as far north as the Euphrates. The ancient Arabs were wont to bind its long curved horns together so they grew as one; and in this form, the animal gave rise to the myth of the unicorn, as recorded by Aristotle. It is marvellously adapted to the desert – better even than a camel, for it does not need to drink at all except sometimes when pregnant, or when the desert grazing is particularly parched. It can, however, gather moisture from the cool humid winds that blow in at dawn from the Arabian Sea, to condense on rocks or on each others' hair.

The behaviour of the Arabian oryx is also adapted beautifully to the desert. It tends to walk rather than run. If it needs to move to new grazing, it travels mainly at night. Individual animals have a truly prodigious sense of direction – apparently based on an ability to map their surroundings: so if they do need to drink, for instance, then they can move unerringly to watering places remembered from some previous journey. They tend to stay together in herds, and if they become isolated, they stand on high places so that they can easily be seen by their fellows. Their white coats make them extremely conspicuous.

But this behaviour – walking, herding, making themselves visible – also makes them easy to hunt. The Arabs have indeed hunted them for thousands of years. But in the mid-nineteenth century the Arabs acquired rifles, and in the mid-twentieth the oil industry brought vehicles with four-wheel drive, and after that, the Arabian oryx was ridiculously

easy meat. They have been in decline since the nineteenth century, and the last one in the wild was reported killed in 1972.

Fortunately, however, the Fauna Preservation Society perceived the oryx's decline and in the spring of 1962 they managed to capture four from the wild (of which one died); and then gathered one more from London Zoo, and four from King Saud of Saudi Arabia, to establish the so-called 'world herd' in Phoenix Zoo, Arizona.

Phoenix Zoo did its job extremely well. With this tiny nucleus of animals, plus a few others that came to light in private parks in the Middle East as time passed, they bred the animals and distributed the offspring first to other American zoos and then to European zoos, until the captive herd outside Arabia was numbered in scores. In the 1970s the Sultan of Oman made known that he would like the oryx to live in his country once more as a wild animal; and pledged that his people, the Harasis, would look after them. In 1980, then, under the direction of the English zoologist Dr Mark Stanley Price, two small herds were taken into a holding enclosure at Yalooni. At intervals afterwards, the two herds were released.

The reintroduction was far from easy. The animals had to get to know each other and establish

a stable hierarchy, with strong intelligent leaders, before they were finally released into the wild. Some animals were not allowed into the wild, because they were too friendly towards human beings. After release, the animals had to get to know their terrain. They made mistakes at first, and wandered into impossibly hostile places; but built up their knowledge of a wider and wider area as the years passed. It was essential, too, that the local people should welcome the newcomers, for without their co-operation reintroduction was doomed. The Harasis have in fact been excellent wardens.

Now there are at least 400 Arabian oryx worldwide, of which about 300 are still in North America. Mark Stanley Price is still worried that the ones in the wild may yet run into genetic problems caused by inbreeding, partly because the total numbers are still small, but partly, too, because only a minority of animals are actually breeding. However, as the wild herd breaks up into smaller populations, this problem may be solving itself. Overall, then, it seems that the salvation of the Arabian oryx has been a success: that captive breeding and reintroduction is a viable strategy.

Now, indeed, about one hundred such reintroductions are in operation, or are planned, worldwide. Others include the black-footed ferret into Wyoming, and the red wolf into Carolina. Mark Stanley Price warns, however, that not all reintroductions can be as straightforward as the Arabian oryx – and that has been tricky enough. The desert, after all, is a relatively 'simple' environment, at least for animals adapted to it. It is two-dimensional, contains no big predators these days, and has only a small variety of plants, nearly all of which are edible. Contrast this with tropical forest, which is three-dimensional, full of hazards, and contains a bewildering variety of species, only a few of which are truly nutritious, and these appearing in unpredictable places and at irregular intervals. It is for such reasons that reintroduction of the orang-utan into Sumatra, which has been tried many times in recent years, has not so far succeeded.

Difficult does not mean hopeless, however. Reintroductions of the golden lion tamarin into Brazil have been successful, and this is a tropical forest species. For more and more species, captive breeding with reintroduction probably offers the best hope of salvation.

▼ *Arabian oryx have adapted to the desert partly by being conspicuous – which helps them to stay in touch with each other. But it also made them extremely easy to shoot.*

▲ ◄ *Ways must be found to enable people to see animals in the wild, without disturbing them. Forest biologists in Queensland dream of walk-ways through the canopy, where tourists can watch the lizards and the tree kangaroos (above). Left shows how not to do it; droves of trippers harass a beleaguered leopard in Kenya.*

Wildlife in cities

Again, the idea that wild creatures could live in cities – and still be genuinely wild – is also new. Just a few years ago it was generally considered ludicrous. Birds were banned wherever possible from city centres (because they were considered noisy and dirty); and municipal parks were above all tidy, with cultivated plants and trees, manicured lawns, and wild-flowers rooted out as 'weeds'.

Attitudes are changing, however. In Salt Lake City and in Prague, Peregrine falcons are being encouraged to nest. The urban kestrel is a common and welcome site in London – as is the heron. Along the New England coast (not quite urban, but certainly heavily populated) and in Florida, ospreys are encouraged to nest on poles in the marinas. Canada geese have greatly enriched the London landscape in recent years. Londoners are no longer quite so neurotic about foxes, which are common in the suburbs. Owls are encouraged. The greater-spotted woodpecker is becoming positively commonplace. In Holland, housing estates these days are commonly built amid natural forest; and there and in many other places in Europe, natural forest is being reintroduced into land that once was a semi-formal (and infinitely tedious) garden.

Perhaps, one day, we may look forward to architecture that is designed specifically with wildlife in mind; houses with cellars built

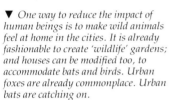

▼ *One way to reduce the impact of human beings is to make wild animals feel at home in the cities. It is already fashionable to create 'wildlife' gardens; and houses can be modified too, to accommodate bats and birds. Urban foxes are already commonplace. Urban bats are catching on.*

for foxes and badgers, and with gables pierced to admit bats. There are theoretical dangers in all this, including the possible spread of diseases. But they are theoretical. The risks are small. And the reward – a kind of modern Arcadia – would be very great indeed.

Even at this late hour, then, it is probably possible to save the human species from a catastrophic collapse. It is possible, too, to save at least a reasonable proportion of our fellow creatures. If we are to do so, though, we have to address the problems seriously. It is not enough to pay lip service to them, as politicians are so apt to do. What, then, is required?

What is needed

The problems of survival, ours and our fellow creatures', require knowledge and technique. We may learn from the past, of course, but it is no solution simply to re-create the landscapes and methods of our ancestors. Their techniques could not feed the present world population, and would not begin to cope with modern conservation. Agriculture that is productive enough to feed us but is also 'benign' is extremely difficult to devise; the ecology of national parks is extremely delicate; captive breeding and reintroduction are full of pitfalls. We need good science, in short, to do the things that are needed.

We must ask, too, whether present economic and political systems truly address the deep problems of the planet. Conventional politicians express 'environmentalist' sentiments. Yet the economics both of the various manifestations of socialism, and of capitalism, are rooted in the notion that the resources of the planet should be seen as raw materials; and that 'success' can reasonably be assessed – or indeed *must* be assessed – in terms of Gross National Product (GNP), and expressed simply in cash terms. Conventional economics of the kind discussed in present parliaments does not include the essential concept of sustainability, which has become the crucial issue. We need new ways of costing labour and resources, and of judging the quality of life. Present-day political arguments are absorbing, and cause much passion. But if all policies lead ultimately to the degradation of the planet, it really makes no difference which we choose. There are, now, institutes of Green economics, where the necessary principles are being worked out. We will need to act upon them as soon as possible.

The underlying essential is attitude. We have been encouraged in the belief that it is possible and indeed desirable to strive for ever-increasing material comfort. But that is simply not the case. We have all of us in the western world been told, particularly in the past decade, that the fabric of the planet is there to be exploited, by whoever is prepared to pay the most. That is simply naïve. We have been brought up to believe that we are superior to all the rest of Creation and have a right to dispose of it as we will. In a previous age, such an attitude would have been considered blasphemous; and so it is.

If we want to survive, as we surely do, then the only proper attitude to this planet and to the other creatures that live in it, is one of reverence. One function of science is to help us to control the world we live in. The other, which in the end should be seen as the more important, is simply to help us to appreciate it.

Glossary

Amino acid Amino acids are the basic 'building blocks' from which proteins are constructed. About 20 different kinds occur commonly in nature, but all of them are organic molecules containing hydrogen, oxygen, and nitrogen (in addition to carbon), and a few also contain sulphur.

Assemblage A group of creatures which may or may not be of different species, which are all gathered together in one place but are not necessarily interacting with each other.

Autotroph Autotroph means 'self-feeding' (in Greek). Plants are nature's principal autotrophs because they 'feed themselves' by capturing energy from the Sun (by photosynthesis), carbon dioxide gas from the air, and inorganic nutrients (such as nitrate) from the soil. Some photosynthetic bacteria are also autotrophic.

Biomass The total mass of material derived from living things in a given space or sample.

Biome A kind of habitat, considered on the grand scale; such as tundra or savannah.

Carbohydrate A sugar, or any compound made by joining sugars together – such as starch, glycogen, or cellulose.

Carnivore An organism that derives all or most of its diet from eating animal material.

Catalyst A substance that aids or accelerates a chemical reaction without itself undergoing change.

Chloroplast A structure (see organelle) within a plant cell which contains the green pigment, chlorophyll, and other pigments involved in photosynthesis. Chloroplasts also contain a small amont of DNA.

Community A group of organisms that live alongside one another, and in which the different species and individuals interact with each other.

Convergent evolution When two kinds of organism are exposed to similar environmental pressures and separately evolve structures or behaviour that are similar. Thus kangaroos, which are marsupials, and ruminants such as cattle and deer which are eutherians, separately evolved big stomachs full of bacteria for fermenting bulky plant food.

Cosmic rays Charged particles bombarding the Earth from space.

Cytoplasm The 'ground substance' of a cell. It is commonly described as 'jelly-like'. But it is clear now that cytoplasm has a tightly controlled and intricate structure, interlaced by membranes. The cell's various organelles are suspended within the cytoplasm.

DNA See nucleic acid

Ecology The study of the interactions between organisms and their physical environment, and between different kinds of organism.

Ecosystem A word that embraces both a particular kind of habitat, and the interactions of all the creatures within it. For example, an oak tree with all the creatures that live within it is an ecosystem; the leaf litter in an oak wood is an ecosystem; and an oak forest is an ecosystem, which includes both those smaller ecosystems, and several others.

Electromagnetic radiation The energy given off when an electron changes its position (or in some other way alters its energy state). Electromagnetic radiation is sometimes described most conveniently as if it consisted of a stream of energy particles (known as photons) and sometimes is best thought of as 'waves'. The shorter the wavelength (as in gamma rays and X-rays) the higher the energy. The most important form of electromagnetic energy for living things is light.

Element The simplest kinds of chemical substance as recognized by conventional chemists. There are about 100 different kinds of element, roughly divided into metals and non-metals.

Enzyme A protein, sometimes with other materials attached (or in close attendance) that acts as a biological catalyst. Most of the body's chemistry – its metabolism – is controlled by enzymes, so the kinds of enzymes an organism has determine how its body is structured, and how it works.

Eukaryote Greek for 'proper cell'. In the cells of eukaryotic organisms – which means of plants, animals, and fungi – the DNA is contained within a distinct nucleus, separated from the cytoplasm by membrane.

Food chain A succession of creatures, each one consuming the one below. Thus plants are eaten by animals which might be eaten by larger animals, and so on.

Food web Most food chains are branched: eg, most plants are eaten by more than one kind of animal, and most animals eat more than one kind of plant (or more than one kind of other animal). Thus, in nature, the pattern of who-eats-whom is usually more like a web than a simple chain.

Gene Genes are the 'units of heredity'; the 'things' that are passed on from parent to offspring and determine the form and function of the offspring – or at least define the range of forms and functions that offspring can achieve. In eukaryotes, bacteria, and most viruses, genes are constructed from DNA; though in some viruses, the genes are made of RNA. Genes function by producing proteins, many of which are enzymes, and which therefore determine how the organism functions.

Herbivore A creature that derives all or virtually all of its diet from eating plants.

Heterotroph Greek for 'different feeding'. An organism that survives by consuming organic material that has already been created. All animals, fungi and most bacteria are hetrotrophs, and a few plants (such as the insectivorous plants) practise some degree of heterotrophy.

Hormone A 'chemical messenger'; a chemical produced in one part of the body that influences the behaviour/metabolism of another part of the body.

Insectivore A creature that derives all or most of its diet from eating insects.

K-strategy Creatures that produce only a few offspring but also supply some measure of protection (which may or may not include parental care) so that a high proportion of the offspring survive, is a K-strategist. Contrast this with r-strategist.

Lineage A line of descent; successive generations of creatures all descended from a common ancestor.

Metabolism The total chemical activity of the body, or of a part of the body, or of the cell.

Mitochondrion (plural mitochondria) Mitochondria are structures (organelles) within the cell that contain the enzymes responsible for respiration; sometimes known as the 'power houses' of the cell. Mitochondria also contain some DNA which can have important genetic functions.

Natural selection The evolutionary process proposed by Charles Darwin. The basic idea is that all creatures produce more offspring than their environment can support; those offspring are therefore inevitably in competition, both with each other and with other creatures; only a proportion can therefore survive; and those that do survive are the ones that are best adapted – or, as Darwin said, are the 'fittest' (in the sense of 'most apt').

Nitrogen fixation The process by which some bacteria are able to capture nitrogen gas (N_2) from the air, and 'reduce' it (which means combining it with hydrogen) to form ammonia (NH_3). Plants are able to take up the ammonia in ionized form (NH_4^+); or it may be converted by other bacteria into nitrate (NO_3^-) which plants can also absorb. Plants then use the nitrogen in the NH_4^+ or NO_3^- as components for nucleic acids and proteins.

Nucleic acids Nucleic acids are of two kinds: deoxyribonucleic acid, or DNA; or ribonucleic acid, RNA. DNA is the material of which genes are made. RNA comes in various forms, but in general it serves to carry out the 'orders' of the DNA, and assists in various ways in the construction of proteins. Some viruses do not have DNA at all. In them, RNA serves

also as the genes.

Nucleotide Nucleotides are organic molecules containing a 'base' (which in this context is a small ring-shaped molecule containing carbon, hydrogen, oxygen and nitrogen); a sugar (which may be ribose or deoxyribose) and a phosphate group (containing phosphorus). Both DNA and RNA both consist of strings of nucleotides. The sequence of nucleotides corresponds to the sequence of amino acids in the proteins that are produced from them. In other words, the nucleotide sequence provides a code for the proteins; the 'genetic code'.

Nucleus A conspicuous structure (organelle) within the cells of eukaryotes, which contains most of the cell's DNA; that is, most of its genes.

Omnivore A creature that derives its diet from a mixture of animal and plant material.

Optimum foraging strategy A 'strategy' unconsciously adopted by animals which ensures that they gather food efficiently. For example, the intake of an animal such as a lion is a balance between food that may be less nutritious, but is easy to come by; and food that is harder to come by, but more valuable.

Organelle A distinct structure within a cell that carries out some special function. The nucleus, chloroplasts, and mitochondria are examples of organelles.

Organic Any molecule in which carbon is predominant can properly be said to be 'organic'. In practice, most organic molecules on Earth are produced by living things. But there are many organic molecules in the Universe at large which most scientists believe are not derived from living things.

Organism The word organism is commonly used simply to mean 'creature'; any living thing. But the word has more profound connotations, which are discussed on pages 1–2.

Parthenogenesis In some animals and plants, offspring develop from unfertilized eggs. This is parthenogenesis. Thus, parthenogenesis is a form of asexual reproduction which has evolved from devices – eggs – that originally evolved as a component of sexual reproduction.

Pathogen An organism that causes disease.

Photon A 'packet' of electromagnetic energy, which may or may not be a packet of light energy. See electromagnetic radiation.

Phytoplankton Plants living in the plankton.

Plankton The assemblage of organisms that float near the surface of large bodies of water, such as oceans and lakes.

Population 'Population' has various definitions, depending largely on context. Usually, though, it means a group of creatures of the same species which are in sufficiently close contact to enable the different individuals to interbreed.

Predator A creature that preys upon another creature. The word is applied most commonly to carnivorous animals, which eat other animals. But biologists commonly say that herbivorous animals are 'predators' of plants; and there is no reason why the word should not be used in this way.

Prokaryote Greek for 'preliminary cell'. An organism in which the genes are not contained within a nucleus, but instead lie in various structures (including plasmids, and giant chromosomes) within the cytoplasm. The commonest and most important prokaryotes by far are the bacteria.

Protein A large 'macromolecule' consisting of a chain (or several chains) of amino acids (each of which is a 'molecule'). Proteins serve many functions in the body. Many are structural. Some serve as hormones. Many – most – serve as enzymes.

r-strategy Organisms that produce a large number of offspring, but offer them little protection so that a large proportion are bound to perish, are said to practise 'r-strategy'.

Respiration (aerobic and anaerobic) The process by which organic molecules are broken down to provide energy. In **aerobic respiration** oxygen is employed in the breakdown. In **anaerobic respiration** no oxygen is required.

RNA See nucleic acids.

Saprophyte An organism that lives by breaking down organic material, though without consuming it first. Many fungi and bacteria – the principal agents of decay – are nature's saprophytes.

Species In general, a group of organisms that are capable of breeding together by sexual means. For discussion, see page 124.

Trophic level Each stage in a food chain (or web) is called 'trophic level'. Eg, in the chain, plant – rabbit – fox, the plant, rabbit, and the fox each represents a 'trophic level'.

Virus The simplest kind of living organism. Each virus 'particle' consists only of a core of nucleic acid (DNA or RNA) surrounded by a protective protein coat. Viruses are therefore obliged to live as parasites within the cells of bacteria or of eukaryotes, and are the principal agents of disease.

Zooplankton Animals (including eggs) within the plankton.

Picture credits

Index

acid rain, 38–9, 85, 92–3
adaptation, 2–3, 136–8
adenine, 5
agriculture,
  cattle, and methane gas
    production, 86–7, 91
  conservation with intensive
    production, need, 153
  limiting factors removal by
    subsidies 118–19
  nitrogen cycle, 104
  old-fashioned, 151–2
  and population increases, 118,
    119, 147–8, 150
  terraced rice fields, 150
  see also fertilizers
air,
  climate, 27
  landscape formation, 25
algae,
  diatoms, 46–7, 90
  see also plankton, phytoplankton
algal blooms, 30, 37
  and eutrophication, 117
  plankton in oceans, 53
algal symbionts
  in coral reefs, 56–7
  lichens, 139, 140
alkaloids, 6
amino acids, 5, 16–17
ammonia, 7, 15
  atmospheric, 10, 11, 12, 85
  nitrogen cycle, 104–5, 114
Anabaena, 106
anchovies, 28–9, 42
anglerfish, 52, 53
aphids,
  reproductive strategies, 123
Arabian oryx, 66
  captive breeding and
    reintroduction, 156, 158–9
archaebacteria, 15
assemblages, 140–1
  definition, 121, 122
Astronesthes, 52–3
Aswan Dam, 35, 37
atmosphere,
  composition, 10–11, 84–5, 90
  primitive, 11, 93
  radiation filtration, 10, 85, 89
  radiations, and gaseous
    interactions, 12, 85

  see also pollution
Atriplex salt-bushes, 34
Australopithecus, 134
autotrophs, 101
Azolla, 106

bacteria, 14–15
  cellulose digestion by herbivores,
    108, 109
  cellulose digestion by termites,
    66
  in eukaryotic cell evolution, 18
  nitrogen fixation, 101, 104, 106,
    140
  organic matter decomposition,
    64, 114
  pathogenic, 14, 102
  sulphur bacteria, 94–5
  see also archaebacteria;
    cyanobacteria
balance of nature, 129–30, 133
  population size, 126
  upset, and atmospheric
    pollution, 90–2
bats, 78
Beagle HMS, 63
Betula nana dwarf birch, 71
bioluminescence, 44, 53
biomes, 61–2
  fluctuations over time, 129–30
birch, dwarf *Betula nana*, 71
birds,
  boobies, 72, 73, 75
  islands, 72–4
  Lake Ichkeul, 36
  learned adaptations, 136–7
  pelagic zone, 49
  reproductive strategies, and
    population size, 131–2
  tundra, 71–2
  waders, 27, 30
broadleaved trees, 69, 70
burning,
  atmospheric pollution, 87, 90–1
  plant successions, 143

cacti, 66–7
  crassulacean acid metabolism, 96
Cairns-Smith, Graham, 3–4, 15
calcium, 7
calorie, 22n.
CAM (crassulacean acid

  metabolism), 95–6
cancers, 83, 89
carbohydrates, 5, 6
carbon, 11
  basis of organic molecules, 4–6
  extraterrestrial, 15
carbon dioxide,
  atmospheric, 11, 84
  greenhouse effect, 86–7
  sources, 90–1
carbon monoxide, 11
carnivores, 101
  food chain, 110–11
  savannah, 66
  temperate grasslands, 68
  tundra, 71–2
catalysts, 5
  atmosphere, 85, 89, 90
cattle,
  methane gas generation, 86–7, 91
cells, 14, 16
  eukaryotic cell evolution, 17–18
  plants, 18, 21, 22, 108
cellulose, 5
  digestion by herbivores, 108, 109
  fermentation by termites, 66
CFCs, *see* chlorofluorocarbons
chance, and survival, 3
Charles Darwin HMS, 42
chestnut coppice, 77, 78–9
Chlamydia, 131, 132
chlorofluorocarbons (CFCs),
  ozone layer breakdown, 85, 86,
    89, 92
chloroplasts, 18, 95, 96
cholesterol, 5
chromosomes, 14
cilia, 18, 48, 52
cities, 77
  wildlife in 161–2
classification of species, 124–5
clays,
  origin of life, theory, 15
  with properties of life, 3–4, 6
climate, 27, 61, 62
  El Niño current influences, 42
  tropical rain forests, 63
  tundra, 70–1
  see also global warming;
    temperatures
climax vegetation, 143, 144
clones, 123